Representing Landscape
Architecture

Plate 1 [see also 1-1, page 22]
English School, Llanerch,
Denbighshire, c. 1662–1672

Yale Center for British Art,
Paul Mellon Collection, USA

Plate 2 [see also 3-3, page 58]
Walter Hood. Autry National
Center, Los Angeles, California,
2006.

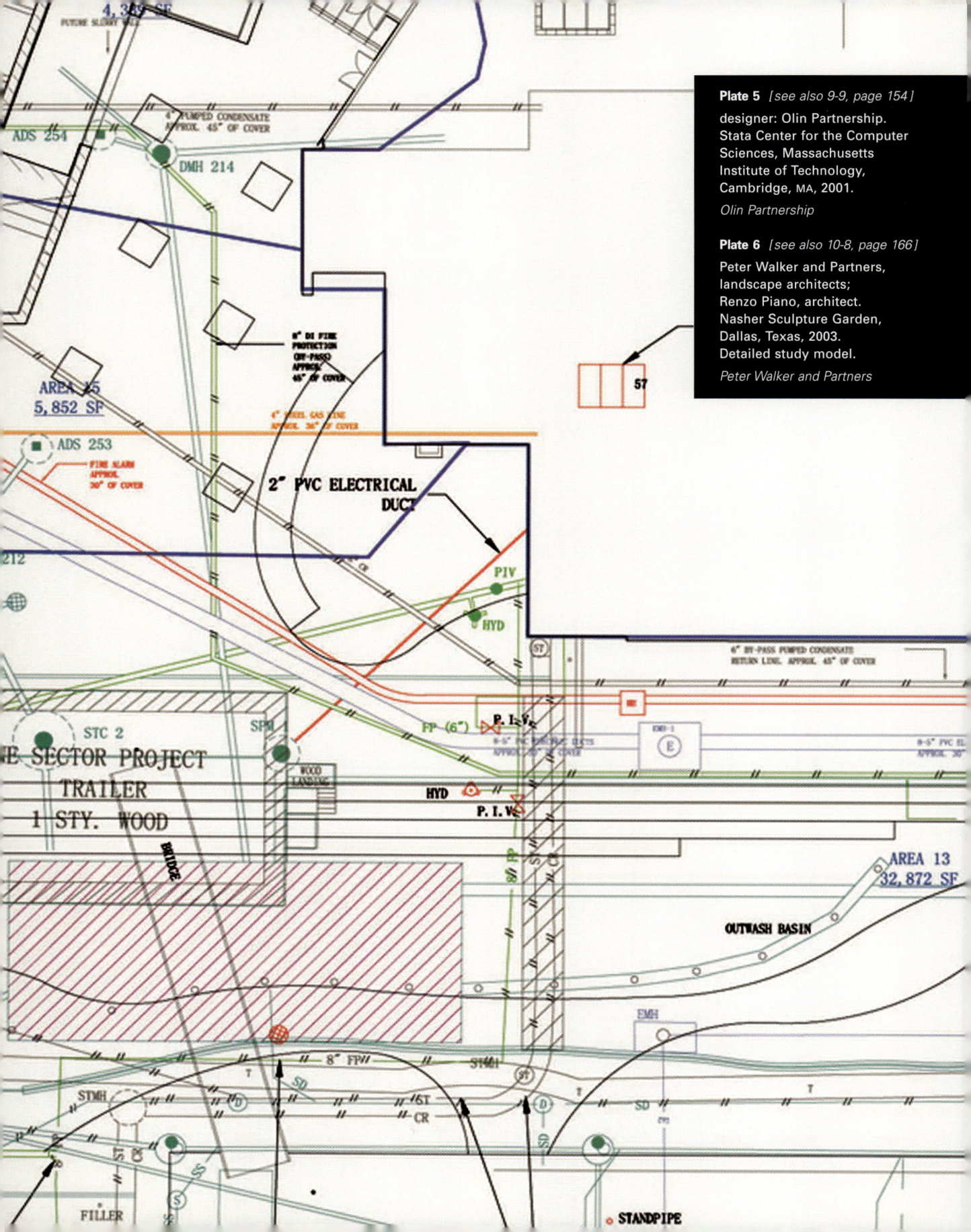

Plate 5 *[see also 9-9, page 154]*

designer: Olin Partnership. Stata Center for the Computer Sciences, Massachusetts Institute of Technology, Cambridge, MA, 2001.

Olin Partnership

Plate 6 *[see also 10-8, page 166]*

Peter Walker and Partners, landscape architects; Renzo Piano, architect. Nasher Sculpture Garden, Dallas, Texas, 2003. Detailed study model.

Peter Walker and Partners

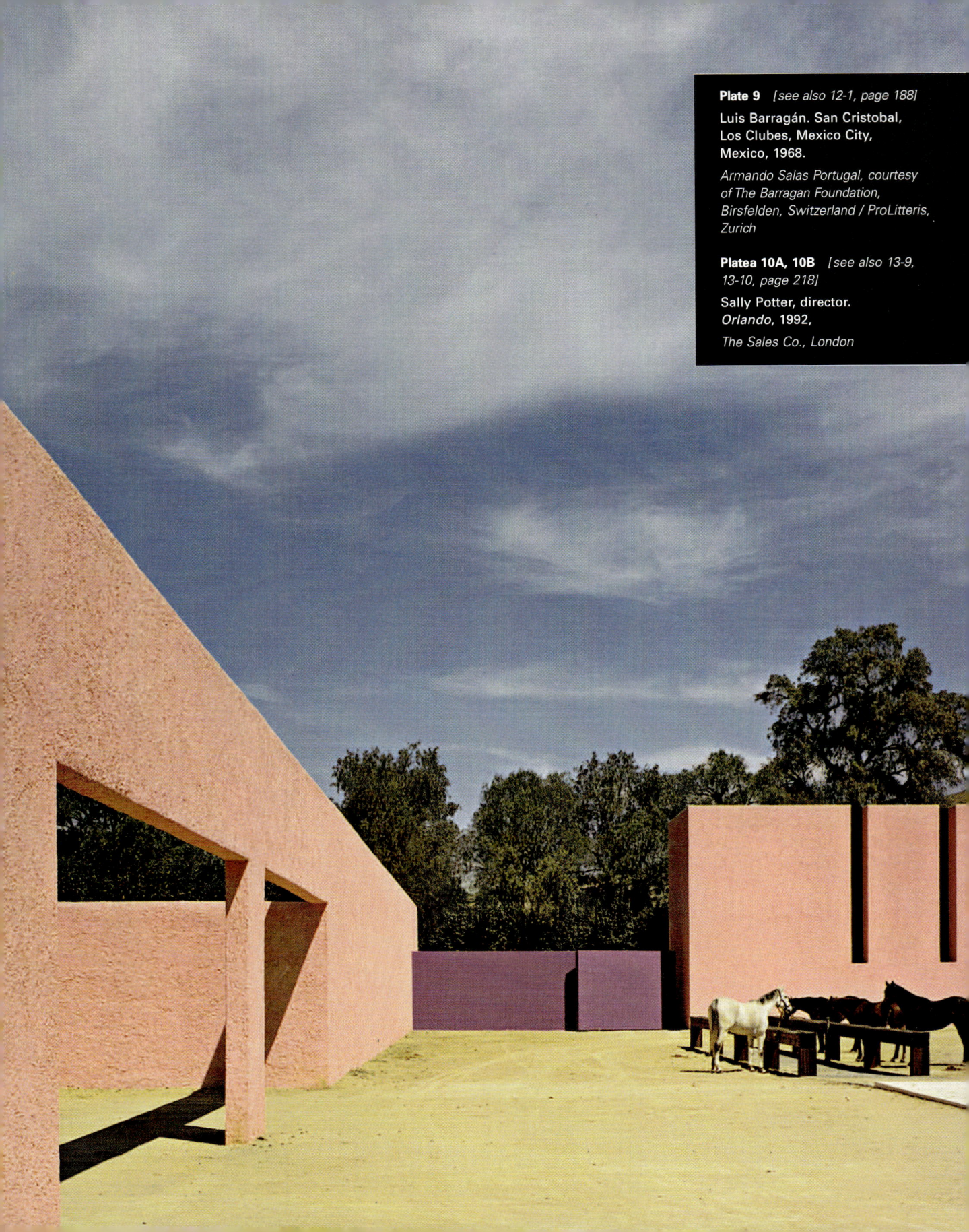

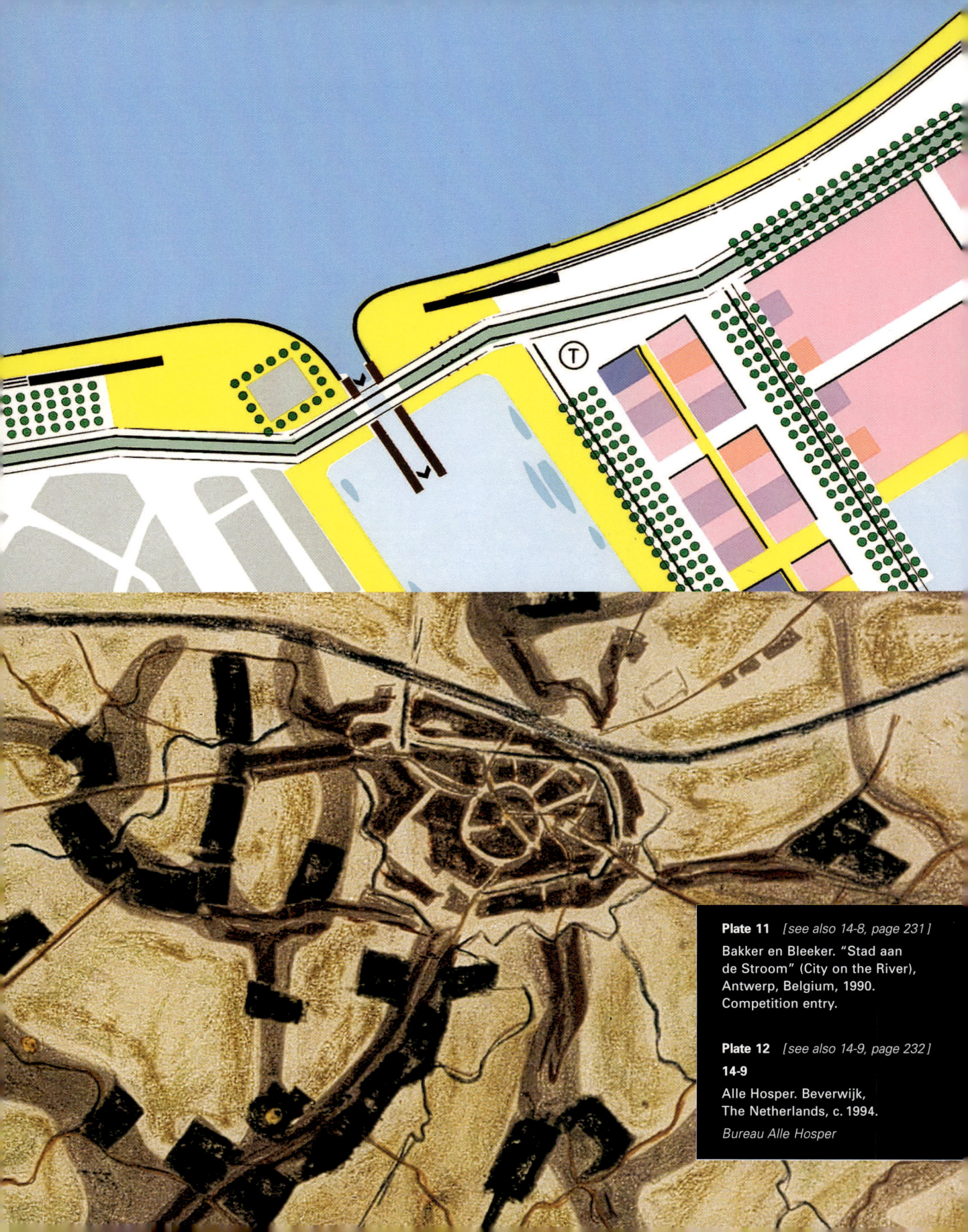

Plate 11 [see also 14-8, page 231]

Bakker en Bleeker. "Stad aan de Stroom" (City on the River), Antwerp, Belgium, 1990. Competition entry.

Plate 12 [see also 14-9, page 232]

14-9

Alle Hosper. Beverwijk, The Netherlands, c. 1994.

Bureau Alle Hosper

Representing Landscape Architecture

Edited by Marc Treib

Taylor & Francis
Taylor & Francis Group

LONDON AND NEW YORK

First published 2008 by Taylor & Francis
2 Park Square, Milton Park, Abingdon, Oxon OX14 4RN

Simultaneously published in the USA and Canada
by Taylor & Francis, 270 Madison Avenue, New York, NY 10016
Taylor & Francis is an imprint of the Taylor & Francis Group,
an informa business

Designed by Marc Treib
Typeset in Matrix Book and Univers
Printed and bound in Great Britain by
The Cromwell Press, Trowbridge, Wiltshire

British Library Cataloging in Publication Data
A catalog record for this book is available from the British Library

Library of Congress Cataloging- in-Publication Data
A catalog record for this book has been applied for
ISBN10 0-415-70042-6 (hbk)
ISBN10 0-415-70043-4 (pbk)

ISBN13 978-0-415-70042-9 (hbk)
ISBN13 978-0-415-70043-6 (pbk)
ISBN13 978-0-203-41281-7 (ebk)

For Nina Hubbs Zurier
For John Zurier

We live in an era of the "new": new technologies, new media, new economies, new life styles, new political and cultural boundaries. In the design professions we repeatedly hear the call for new building and landscape forms, and for new ways by which to conceive and communicate them. We are told that the old forms and old media just no longer work and that, instead, we need new ways of visualizing and presenting design information. This may be true. But very rarely—almost never—do we learn just what is wrong with existing practices and just which of them might need to be revised or replaced. The fourteen essays in this book provide a broad investigation of how landscape architecture has been, is currently, and may be represented in the future: for design study, for presentation, for criticism, and even for its realization.

It has been said that we can only realize what we can imagine. But in order to realize the constructs of our imagination we must convey ideas to others as well as to ourselves. Representation is by no means a neutral practice, and the process of communication, the process by which the imagination takes its first form, itself necessarily limits the range of our design possibilities. Machines such as the computer further remove perceptual from cognitive processes and raise new questions about methods and limits—although, of course, they might augment the power of those processes in other ways.

Is there a link between the media and drawing types we use in creating landscapes, both as promise and limitation? The thought behind this book is that there is still much to be learned from where we have been, especially when projecting where we are going: let us examine, even in a cursory and incomplete way, the ideas and forms by which landscape architecture is and has been graphically represented and described. As the essays will suggest, those forms of representation have responded to a multitude of questions, each one multi-faceted. By whom is the image made and to whom is the communication intended? What graphic media and technologies were available to the designers? What are the possibilities and liabilities for the various media and for various types of graphic representations? How were certain issues concerning landscape design addressed and conveyed in different historical periods? How might ideas from the past lead us to more effective means and manners today and in the future? What is the relation between the representation and the built form? These are the themes and sub-themes that run through the essays, although none of them constitutes a dominating structure. Our concerns include: descriptions of space, form, and vegetation; representations of individual and collective goals; expressive capabilities; issues of time and process and change; lessons from history and how they inform our thinking and representation today; the impact of mechanical and electronic media; representing the construction process; and suggestions concerning the computer and its global extension through the internet.

Many of the "new" forms one currently encounters are hardly new. In some instances they border on decoration, for example, inserting a photograph between two charted lines on a graph or superimposing two images upon one another. Rather than contributing additional rigor or clarity to the graph, the image often undermines its original clarity. The advent of software programs such as Photoshop has granted an enormous power to designers in terms of realism and accuracy, but these may be achieved at the expense of a sense of life and a confusion of detail for idea. Because it is easy to generate numerous computer images once the data have been entered, we encounter floods of pictures rather than one or a handful that might convey the gist of the idea in a more lucid form. There also appears to be less thought given to the purposes and relations among the types of views—the plans, sections, perspectives, axonometrics—not to mention their relation to the possibilities offered by photographs, film, and today animation and video. These combinations lay at the heart of the matter: that one drawing type in isolation is usually insufficient. A plan without a section or elevation may be of little utility and vice versa. A perspective provides pictorial information but not necessarily any insights into how the elements of that view would be realized. Too much imagery today—at least in my view—is produced primarily because we can produce it, and often at the expense of the design idea, the qualities and experience envisioned, and the recipient's ability to decipher the information. Behind everything in this book then is a call for representation to be linked to thinking rather than to the mere creation of special effects that capture the eye without necessarily effectively engaging the brain.

My critical stance on certain "new" graphic representations is neither a call nor excuse for complacency. Certainly in a world of interactive media the static view taken alone may no longer provide the solution to the problem at hand. In addition to the age-old issue of the combination of views characteristic of traditional presentations—plan, section, elevation, axonometric, perspective, and of course diagrams and text—in an age of interactive media we must be more concerned with time than ever before. Animated simulations and video offer enormous possibilities given their ability to create a narrative available until recently only in film. But even the time-based media can employ the traditional views, even if they are now linked in time along a directed path of viewing. To some extent theses issues lie beyond the purview of the current volume and leave open the opportunity for additional study, but much of the thinking in the essays—if not the actual media discussed—are applicable to time-based graphic forms.

Representing Landscape Architecture has developed from the symposium "Representing the Designed Landscape: Images, Models, Words," held at the University of California, Berkeley in 2001, and sponsored by its College of Environmental Design and Department of Landscape Architecture and Environmental Planning. For the most part, the book expands on the papers presented at that conference, although several of the speakers, for various reasons, were unable to contribute to the resulting volume. Nonetheless, their participation should be acknowledged with gratitude: James Corner, Georges Descombes, and Hope Hasbruck. On the other hand, we have added several essays by authors who did not speak at the symposium to examine additional subjects germane to the general topic. Certainly, not all of those subjects have been addressed, alas, and there are sure to be readers whose own personal interests will not be discussed either in part or as a whole. Unfortunately, for example, attempts at commissioning an essay on language as landscape representation remained unsuccessful. But one book can rarely cover a subject exhaustively, especially given the economic restrictions that limited the number of essays, the almost exclusive use of black-and-white images (color is another major issue missing from most of these discussions), and only a representative selection of images for each essay—a significant reduction from the barrage of projected images offered by each speaker at the symposium.

The reader will note that of the many topics in the book, the computer is touched upon only briefly, as an end point and harbinger of new directions. Firstly, computer-generated imagery is such a vast topic that it warrants a study of its own—and it already has received considerable attention, a regard that is continually growing. Secondly, we believe that by better understanding the achievements of more traditional means we can better utilize the capabilities of the new technologies at the disposal of designers today. Thirdly, of course, is the perennially nagging limit of space: this book focuses on what has been, and what is, as a means of looking forward to what can be. Rather than claiming any pretense at being a conclusion, the essays constitute only a beginning, to provoke thought, discussion, and perhaps further study and broader dissemination.

Marc Treib
Berkeley
January 2007

First, thanks to Harrison Fraker, dean of the College of Environmental Design, former landscape department chair Walter Hood, the Farrand Fund, and Tina Gillis at the Townsend Center for the Humanities—all of whom helped finance the original program. Thanks also to Cheryl Barton for her support, Ron Herman and Peter Walker and Partners underwrote the reception at symposium's end. Chip Sullivan should be credited for the splendid drawing on the poster, now used in revised form on the book's cover. Finally, but certainly not least-ly, I need thank Mary Anne Clark in the Department of Landscape Architecture and Environmental Planning for efforts far beyond those of her job description, tasks that covered every aspect of the program from handling publicity and registration to working out the details for the breaks and the reception.

For bringing the book into reality through Taylor & Francis my sincere gratitude goes to Caroline Mallinder, whose support and enthusiasm have been unwavering. Susan Dunsmore took on the challenging task of editing with efficiency and sensitivity the writing by authors of several nationalities. And finally I need thank Katherine Morton for her untiring efforts to attain a high level of quality throughout production.

1

Dianne Harris and David L. Hays

On the Use and Misuse
of Historical Landscape Views

Like designers, landscape historians must constantly wrestle with the ephemeral nature of their subject matter. Landscapes are events. They begin, develop, transform, and eventually come to an end, sometimes leaving little in terms of tangible remains. Historians therefore rely on various other sources of information about landscapes, but there, too, they must navigate considerable uncertainties. For example, texts describing landscapes often contain hyperbole, fantasy, or projection. They are subject to distortion for a broad range of purposes. If words can deceive, images can do so with equal if not greater effect. In most Western cultures, seeing is equated with believing, and a deceptive picture is worth a thousand deceptive words.[1]

How, then, should we consider the images upon which landscape history has depended so strongly, especially for works antedating the invention of photography? For example, should we trust the many prints that seem to show what villas, gardens, and estates looked like at particular moments in time? The answer is: "rarely." Although they are among the source documents most commonly employed by landscape historians, such images seldom portray the accurate appearance of sites. By examining a selection of Italian, French, and English landscape views dating primarily from the seventeenth and eighteenth centuries, we hope to demonstrate various ways in which those images may or may not be useful to historians and designers alike. Although each of the examples deserves—and in some cases has received—book-length treatment, our aim is to point toward the complexities inherent in landscape views broadly speaking and to suggest avenues for future research.

Most art historians, particularly recent graduates schooled in the relativism of post-structuralist theory, would never suppose that such artifacts of visual culture could be reliable in a mimetic sense. Yet, because of the ephemeral nature of their subject, landscape historians have tended to look to artists' renderings for documentary clues. When using such images—whether prints, drawings, paintings, or other types—historians have tended to assume their verity, and scholars still disagree about the extent to which such visual artifacts are reliable source documents.[2] Such assumptions are not particularly surprising since mimesis is deeply rooted in the art historical tradition. They are also to some degree understandable, given the instabilities inherent in landscape and the notion that an image represents a higher and generally more stable ideal form.

1-1
English School, Llanerch, Denbighshire, c. 1662–1672.

Yale Center for British Art, Paul Mellon Collection, USA

As Michael Ann Holly has noted, "past works of art actually work at prefiguring the shape of their subsequent histories" and "representational practices encoded in works of art continue to be encoded in their commentaries."[3] In many cultures, vision is privileged as if it were the primary sense. Many of the artists who produced estate views worked to create credible images filled with meticulously rendered detail, employing cartographic techniques such as aerial perspective to produce compelling depictions of place—a practice Lucia Nuti has called "the will to graphic persuasion through perspective and shadowing."[4]

Credibility and utility were not always synonymous with consistent perspective, however. For example, an anonymous, mid-seventeenth-century painting of a country estate near Llanerch in Denbighshire, Wales, focuses on a striking garden, with an axial sequence of terraces and enclosures extending east from the hilltop house and descending into the valley below [1-1]. The painter took great care in representing detailed elements of the garden as well as the surrounding context, producing a result so rich and convincing that the historian John Harris declared: "There is no rarer document than this in the whole history of garden art in [Britain]."[5] Even so, the overall quality of the painting has been questioned by many—Harris included— because the perspective shifts in several places within the image, most noticeably at the lower right. Based on that distortion, John Dixon Hunt qualified it as "a rather naive painting" showing a "superb garden." Harris complained that the artist "lost control of his perspective whilst portraying the garden descending through its terraces."[6] Those remarks frame shifting perspective as a sign of technical or conceptual naïveté, in this instance, stemming from the culturally remote situation of the artist. But to suggest that the image in question is deficient because it manipulates perspective is to ignore the specific context in which the picture was made. While sometimes a consequence of practical inexperience, shifting perspective could also be adopted deliberately and for specific ends. Like the bird's-eye view, it emerged as an attempt to conflate the advantages of perspective and mapping in a single image, with the objective of creating more meaningful representations of landscape than could be achieved using perspective or mapping alone. In depicting the estate near Llanerch, the artist may have employed shifting perspective knowingly to depict the estate as a seat of cultural and economic strength. The image shows off an ambitious new garden as well as a host of symbols of economic power specific to the Vale of Clwyd. Furthermore, the artist manipulated perspective to create an illusion of symmetry and order between house and garden that close inspection of the painting itself undermines.

A comparison of two contemporary prints depicting a single setting—the Cascades at Liancourt as represented by Aveline in the 1650s and Israël Silvestre in 1655—undermines the idea of such images as credible visual description [1-2, 1-3].[7] In a garden renowned for its displays of water, the

1-2

Pierre Aveline the Elder (1654–1722), view of the Cascades at Liancourt, 1650s.

Special Collections, Frances Loeb Library, Graduate School of Design, Harvard University

1-3

Israël Silvestre (1621–1691), view of the Cascades at Liancourt, 1655.

Rare Book and Manuscript Library of the University of Illinois at Urbana-Champaign

Cascades comprised twenty-two fountain jets and subordinate basin bowls arrayed along an embankment between a large terrace and a lower garden quadrangle of even greater dimensions.[8] Those two areas descended in sequence away from the west side of the château and were garnished with large quadripartite parterres. The images by Aveline and Silvestre both show the correct number of jets with three basin bowls and a collecting pool below each, but they differ significantly in other details. For example, the configuration of the lower parterres differs radically in the two views. Aveline composed each of the four quadrants using plain turf with a single-jet fountain set in a central basin. In contrast, Silvestre depicted the four quadrants with elaborately cut patterns of turf, but only the two quadrants closest to the château include fountains. In Aveline's image, large trees encroach upon the flanks of the terrace above, mid-sized cypress trees mark the corners of each parterre, and generously spaced shrubs line the edges of the turf panels. In Silvestre's image, neat lines of cypresses retain the flanking vegetation and no shrubs appear along the parterres. Both images depict the parterres as descending through a grove of mature trees, but that arrangement may have been only a pictorial device. A bird's-eye view of Liancourt published by Henri Mauperché in 1654 shows the two parterres flanked by extensive ranges of *berceaux* (vaulted arbors).[9]

Such disparities in form and detail may even exist among images produced by a single artist. When the Milanese printmaker Marc'Antonio Dal Re produced views of the Villa Ferrante Villani-Novati in Merate—a setting included in both the 1726 and 1743 editions of his bound volumes of Lombard villa views known as the *Ville di delizie*—he freely used artistic license, adapting aspects of the view to fit the format of each publication. The 1726 image ranks among his largest productions, an assemblage of two copperplate prints, its dimensions 32.75 in. wide by 43 in. long [1-4]. For the 1743 edition, Dal Re prepared a new plate, compressing the entire view to fit within a single copperplate image spanning the two pages of the folio [1-5]. Distortions in the 1726 view, such as the absurdly cramped spaces of the labyrinth and the theater in the lower corners, seem to have resulted from the delineator's inability to handle the complex perspective taken from a viewpoint at the base of the garden. That particular station point was the most reasonable choice for effecting the maximum display. From that location an oblique station point would not have readily conveyed the descent of the terraces whereas a view from the top of the garden would have obscured many of its lower features. The 1743 view eliminated the cramped spaces through more consistent handling of the perspective. In fact, Dal Re depicted the lower third of the garden at twice the width seen in the 1726 view. Six trees line the edge of each side of the fifth terrace in the earlier print, whereas

1-4

Marc'Antonio Dal Re, "Veduta Generale del Palazzo, e Giardini di Merate del Sig. Marchese Ferrante Novati," *Ville di delizie o siano palaggi camparecci nello Stato di Milano*, 1726.

Dumbarton Oaks, Research Library, Washington, DC

1-5

Marc'Antonio Dal Re, "Veduta del Palazzo, e Giardini del Sig. Marches Novati in Merate," *Ville di delizie o siano palaggi camparecci nello Stato di Milano*, 1743.

Rare Book and Manuscript Library of the University of Illinois at Urbana-Champaign

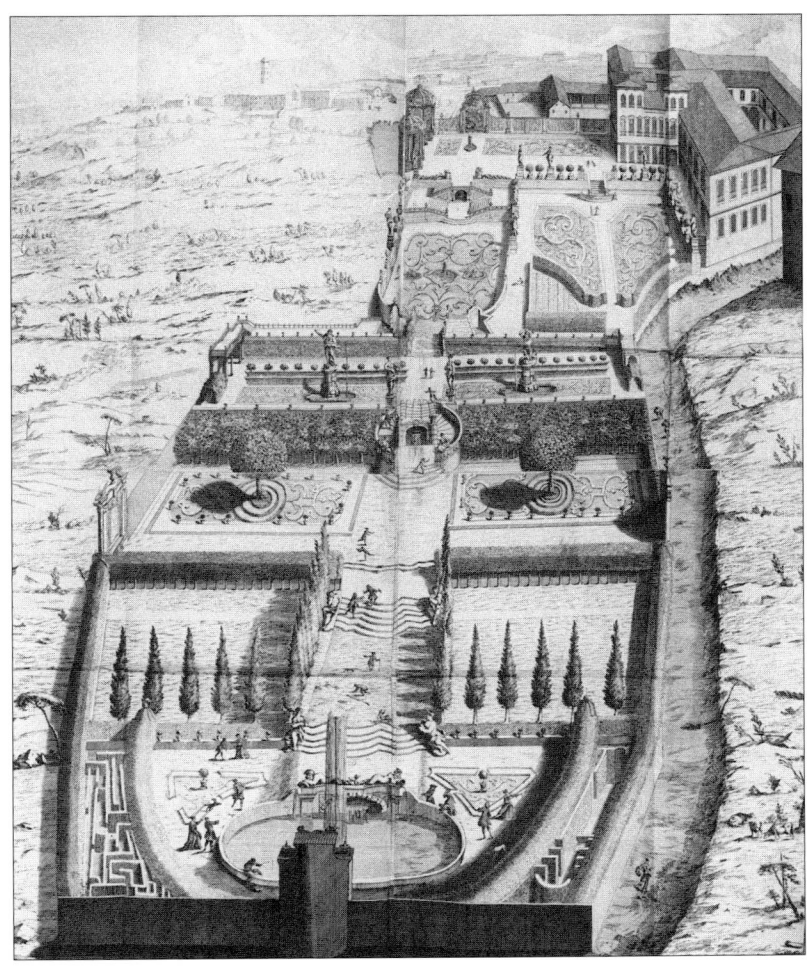

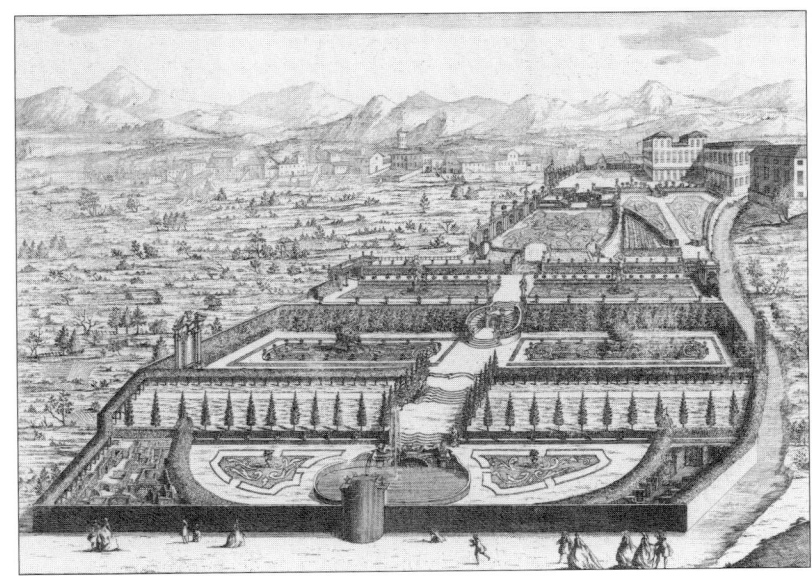

twelve are shown in the same location in the later edition. Today, it is impossible to determine which version is more accurate; the garden was almost entirely destroyed and remade at the end of the eighteenth century, and the wider view may well have resulted from representational convenience rather than from a desire for increased accuracy.[10]

Despite such bedeviling uncertainties, historic landscape views can be highly informative if we look to them for evidence beyond formal and material configurations. Although estate prints may or may not depict the true condition of a particular site, they typically present an image desired by the patron or artist in a format that was easily reproduced and circulated. Although guests at a villa might readily have discerned its owner's status through examination of the architecture, furnishings, and gardens, such visitors were necessarily limited in number. Moreover, the seasonal nature of gardens meant that they were not always in peak condition when guests arrived, particularly during the winter months. Early modern garden, villa, and estate prints represented specific moments in the life of a landscape, presenting idealized and often highly manipulated visions of elite life to an audience larger than the one admitted into that world. Prints allowed the dissemination of images beyond provincial and territorial borders, simultaneously asserting and confirming family status and prestige before local, regional, and international audiences.

Three frameworks for the examination of historic landscape images can help to clarify these points: (1) communication; (2) reception; and (3) perception. Using examples from France, Italy, and England, the principles and questions addressed here may also be applied to images produced in a broader range of cultural contexts.[11]

COMMUNICATION

The first framework for consideration is communication, which begs questions inextricably linked to intention. Obviously, historic landscape views were meant to communicate, but in most instances they did not convey the exact form of gardens in their time. Unlike landscape representations made in design offices today, the estate prints and paintings referred to here were not created to guide a construction process, to persuade a client to hire a designer, or to convince the client to pursue a particular design. In fact, they were not made to persuade the client of anything at all, since they were generally produced at the estate owner's request after a garden was already in place or, at times, even during its installation.

From the Middle Ages through the early modern period, social formation in Europe was based on seigniorial systems of land ownership and management, and two factors prevailed in conferring and maintaining elite status: birth and the possession of a rural estate.[12] Between the sixteenth and eighteenth centuries, however, the criteria of elite identity became increasingly culturally based, with status understood as a function of appearance. As a backdrop for the display of codified civility, distinguished conduct, and genteel dress, domestic architecture and gardens became critical agents in the expression and establishment of social prestige. With a constellation of other signifiers included in views of villas and estates, gardens became part of a visual language for the representation of status. Critical to images of gardens were elements that conveyed the prestige of their patrons and users through display of the most recent styles in garden form, architecture, costume, and comportment.

Although frequently ignored as staffage elements positioned according to artistic conventions, the human figures in landscape views can play very important roles, and their arrangement is usually highly significant.[13] Human figures help establish a sense of scale, and, as such, they can be used to exaggerate as well as to reduce the viewer's perception of space. Figures delineated as if pointing to something specific—for example, toward a significant estate feature—direct the viewer's gaze. Such figures are especially useful for calling attention to elements difficult to represent in views. For example, a fish pond is difficult to identify as such unless we are shown a figure engaged in the activity of fishing, and well-stocked fish ponds were luxury amenities that became potent social signifiers.[14]

In certain images, the property owner appears in the foreground positioned on an elevated vantage point that may or may not have existed in reality. Those features lent authority to the prospect in part because the patrons themselves appear to be looking down on their grounds from an elevated position, in keeping with their social status. In such images, the owners are frequently accompanied by people or artifacts that underscore their elite status. For example, horses, hounds, and other attributes of the hunt signal the possession of hunting rights and preserves [1-6]. The presence of one or two servants alludes to larger retinues. Images of visitors crowding a garden or pressing at its gates suggest the social or political currency of the patron. Israël Silvestre's views of Vaux-le-Vicomte, for example, are particularly exemplary in that regard. In his prospect of the entrance to the estate, an almost impossible swirl of people, horses, and carriages presses for access.

The dress, activities, and placement of imagined visitors also reveal much about the image desired by the artist or patron. Through the careful delineation of clothing and its display on bodies utilizing postures and gestures codified in theater, etiquette manuals, and treatises on the visual arts, artists conveyed complex messages concerning the status of the estate owners and their invited guests.[15] Such images served as indexes of class distinction, making rank legible to an audience of collectors and other viewers. As visual inventories of land-based amenities, the views glorified and confirmed an established elite, whose power and control derived in part from the authority of taste and the projection of specific cultural values and social hierarchies. In that regard, the views present information useful to historians by offering vivid depictions of a social landscape as landed elites wanted it to be perceived.

Of course, the placement, configuration, and identities of figures varied significantly according to the social, political and economic circumstances of each commission.[16] Carmontelle's views of the Jardin de Monceau are especially interesting in terms of the uses to which figures are put [1-7]. Between 1771 and 1779, Louis Carrogis, known to his contemporaries and after as Carmontelle, developed the garden for his patron, Louis-Philippe-Joseph d'Orléans, duc de Chartres, on a site just beyond the northwest edge of Paris. To represent that work, Carmontelle produced a large folio— published in 1779—that coordinated theoretical and descriptive texts, a plan, and seventeen views, the latter keyed to the plan and descriptions.[17] Those elements were organized around an imaginary tour of the garden, a familiar conceit that staged understanding as a function of time. The views are garnished with figures of imaginary visitors, and the viewer is meant to

1-6

Marc'Antonio Dal Re (1697–1766), detail of "Veduta Generale del Palazzo e Giardini di Cinisello del Sig. Conte Donato Silva," *Ville di delizie*, 1726.

Dumbarton Oaks, Research Library, Washington, DC

understand him- or herself as part of that elite company. The costumes of the visitors are stylish and contemporary. They signal that the time of the views is the immediate present—that is, the period of the 1770s during which the garden was realized and the images of it were produced. Significantly, when a later edition of the views was published, the original figures were rubbed out and replaced by figures in updated costumes.[18] Ironically, various settings in the garden had also changed by then, but those transformations were not represented in the new edition.[19]

Curiously, some of the figures within the views of Monceau are shown in exotic costumes (see 1-7). As Carmontelle explained in the text of his folio,

those were servants, and their appearance signaled that the duc de Chartres was then in residence.[20] In other words, the conceit was a social compliment to the viewers of the prints, suggesting that they were in the company of the patron, one of France's most prestigious aristocrats. Also curiously, few of the visitors shown in the views appear to be interested in the settings represented. Instead, they stroll and chat. In that way, Carmontelle suggested that the physical elements of the garden are secondary in importance to the sociability that might take place there.[21]

The question of time in Carmontelle's views is doubly provocative when one imagines them against the probable state of the real garden as a site then under construction. Carmontelle began to prepare drawings for the folio publication long before his supervision of the garden was complete.

1-7
Louis Carrogis, known as Carmontelle, "View of the Farm, Taken from Point D, Close to the Cabaret," *Jardin de Monceau, Près de Paris, Appartenant à Son Altesse Sérénissime Monseigneur Le Duc de Chartres*, 1779, plate V.

Dumbarton Oaks, Research Library, Washington, DC

In November 1776, an entry in the *Correspondance littéraire* mentioned a set of "very varied and very picturesque" images "which only the artist's magic could have produced."[22] The author further reported that the images demonstrated "more invention than in all of our wise theories," but nothing was said about their faithfulness to work being realized on site.[23]

The gardens that appeared as the central subjects of such representations also functioned as reflections of cultural capital based on the international and local cultures of education, literacy, collecting, and theater. Dal Re's prints, for example, are a compendium of ideas considered fashionable for an eighteenth-century Lombard garden, a catalog of garden forms and elements his patrons considered pleasing and status-enhancing. Garden prints, with the written descriptions that often accompanied them, combined artistic and ekphrastic traditions to formulate a total narrative, a discourse on family identity and position. Likewise, choosing the right garden forms, reading the correct theoretical and horticultural texts, and cultivating the requisite number of exotic plants were as important as the collection of material objects in asserting status, and the prints displayed those choices, whether factual or desired. What such prints communicate is the visual syntax of prestige. For that reason, they must be seen as part of the cultural institutions of print. Like other communications media, views of private landscape contributed "to the construction of social reality as a part of the material forces that help to produce and reproduce the world."[24]

1-8

Marc'Antonio Dal Re, "Panorama of the Villa Archinti in Robecco sul Naviglio," *Ville di delizie*, 1726.

The context of the room indicates the size and theatrical quality of the unfolded print.

Dumbarton Oaks, Research Library, Washington, DC

RECEPTION

By whom and in what ways garden images were received—especially in the period of their original production—is often very difficult for historians of visual culture to ascertain. That is particularly true for the historians who study landscape prints, because such images fall outside the art historical canon. The centuries-old tradition of art historical scholarship that privileges, first, painting, then sculpture, and then drawing (in that order) has largely neglected the significance of the print, so that the views that are of such interest to landscape historians are essentially the flotsam and jetsam of the art historical world. For example, such images do not appear in the 1999 Oxford edition of Malcolm Andrews's book *Landscape and Western Art*.[25] Little scholarly analysis exists for the majority of the best-known images in garden and landscape history. Moreover, those images were not part of a salon culture, nor were they exhibited in galleries, so no contemporary published criticism exists.

In some instances, estate images were created specifically to be viewed at close range, as was the case with the panoramic prints of eighteenth-century Lombard villas created and published by Dal Re. In scholarly literature, those images are frequently reproduced in isolation, hovering above some text that uses them to illustrate a garden or estate description. In that context, the images are assumed to document the way Lombard villas appeared in the eighteenth century. Figure 1-8 shows a panoramic print as Dal Re intended it be viewed: folded out on a table. Seeing the prints' true format and practical viewing context does not reveal anything about their accuracy as documentation of realized forms. However, the context for viewing underscores that they are large and very theatrical. The images in

question had to be placed on a table and unfolded, where the details of each view could be carefully examined up close. Unfolding Dal Re's panoramas is itself a theatrical experience, necessitating movements back and forth as the arm stretches to unfold the image. It is an experience not unlike lifting the flaps on one of the watercolored before-and-after views in one of Humphry Repton's Red Books. Both elicit a sense of surprise, delight, and theatrical drama by employing modes of presentation that are engaging and persuasive. Dal Re's printed villa views were intended to be studied at close range, even through magnification, and they tend to be filled with layers of carefully constructed detail that reveal a great deal about the ways patrons wanted to be seen. Little is known about the identities of those who viewed Dal Re's prints and less still about the ways they received them, but understanding the intended viewing context—the fact that they were meant to be examined at close range, their details pored over and carefully studied—helps us understand why every aspect of the images received the careful attention of the artist and must be presumed to be at least symbolically significant.

So far, virtually no evidence has come to light concerning the reception by contemporaries of these printed views.[26] Still, the large numbers of such publications that appeared in the seventeenth and eighteenth centuries indicate a marked degree of popularity, at least within elite circles, leading us to assume that they attracted a substantial audience. Moreover, the production of estate views must have seemed a potentially lucrative venture, with a level of demand that could compensate for their time-consuming and costly production.[27] In addition to the patrons for whom such prints were made, Grand Tourists collected them, purchasing either entire volumes or individual images, and they undoubtedly showed them to friends upon their return home, like so many outsized souvenir postcards.

Prints were also collected by social peers and neighbors, fellow estate owners throughout Europe who wished to emulate and compete in the display of taste and elite culture under a variety of circumstances. As Tracy Ehrlich and Erik Neil have shown, for example, villa owners in both Frascati and Bagheria competed to construct the grandest and most elaborate villas.[28] In both of those cases, the proximity of villa owners allowed them merely to look outside their doors or windows to see what neighbors were constructing. But for those farther afield, bound volumes consolidated printed views of the most fashionable and recent constructions. The fact that such prints were assembled into books made them a part of literate culture. As Chandra Mukerji has noted, "In a culture in which the trappings of literacy were signs of high standing, these highly literate gardens were not just interestingly reflexive; they were appropriate means of claiming rank," and such publications "tied garden design through the theories they referred to with a highly developed literate elite culture."[29]

PERCEPTION

How, then, were such images constructed for an audience of their time? As John Berger wrote in 1972, "Every image embodies a way of seeing," and in the past decade alone numerous publications have appeared that focus on the social construction of vision.[30] Perception derives from a combination of the physiological phenomena associated with sight and the cultural forces that shape the consciousness of the observer. Seeing begins when we open our eyes. Looking is much more selective. Choices surrounding representation are frequently based on presumptions about viewers and how they will look at an image. Even more significant, they are based on presumptions about what the viewer is likely to believe when he or she looks at the representation.

During the seventeenth and eighteenth centuries, estates were often portrayed through elevated perspectives. The frequency with which that format was employed was not merely a result of practical ambitions to conflate the virtues of picture and plan into a single image, though that was certainly a significant factor.[31] Put bluntly, plan views rarely seem as persuasive as other forms of representation, and methods for representing elevational and material changes (e.g., surface curvatures, tall growth versus short growth) were still being developed in the late eighteenth century. The flat views of elevations are even less compelling as images. In contrast, perspective affords a wide range of possibilities and for centuries (at least since Alberti) seemed to correspond to a Western European sense of spatial reality.

The literature on perspective is vast, but the specific point here is that its use in garden prints stemmed from an understanding on the part of artists that such views instilled a sense of credibility.[32] All artists understood the persuasive qualities of perspective views, and it is no coincidence that theater designers from Palladio to the Bibiena family exploited perspective in the construction of stage sets. The drama of scenery that seemed to recede into space lent credibility to dramatic productions and helped audiences immerse themselves in the action on stage.[33] After acting itself, receding perspectives were the primary means of suspending disbelief and of drawing the theater audience into a temporary reality. Garden prints were likewise produced for an aristocratic European culture steeped in the logic of perspective and theatrical habits of perception, and viewing the world through the lens of the theater constituted what Michael Baxandall has referred to as a "period eye."[34] That frame of reference sometimes exerted an impact on gardens in surprisingly literal ways. In Cardinal Richelieu's garden at Rueil, the Triumphal Arch (c. 1638) was, in fact, a *trompe l'oeil* confection painted on a garden wall.[35] In various festival entertainments at Versailles, such as *Les Plaisirs de l'Île Enchantée*, performed in May 1664, the boundary between garden and theater was completely blurred [1-9]. In the theater at Chantilly, constructed a century later, the back of the stage could be opened to reveal a view of a real fountain set into a nearby garden wall.[36]

Among the various practices of perspective, the elevated prospect signaled a commanding position, one that marshaled the administrative metaphor of surveillance and detachment over an extensive domain and combined it with the authority of divinity implied by views from on high before the possibility of mechanical flight. The producers of estate views clearly understood those associations. Giovanni Battista Falda included both plans and elevated prospects in his seventeenth-century bound volumes; Dal Re used elevations and sections, but he always inserted fold-out panoramas as seen from elevated viewpoints as the final statement on the appearance of each estate.[37] Giuseppe Zocchi and Gianfrancesco Costa abandoned plans and relied exclusively on perspective views to tell their stories of eighteenth-century villa life in Tuscany and along the Brenta River.[38] However, prospects did not go unquestioned, nor were they universally embraced.[39] In France, high-angled prospects were especially popular during the sixteenth century but subsequently fell out of fashion. In Great Britain, the format flourished between around 1670 and 1730 but was otherwise rarely adopted. In late eighteenth-century France, the

elevated perspective appears to have been rejected outright by advocates of irregular design because that format alienated viewers from the sense of immersive experience upon which the success of the new mode was thought to depend.[40]

In scholarship related to garden and landscape history, then, each printed or painted view must be examined within its particular cultural context using multi-source analysis.[41] The elevated perspective certainly conveys a sense of credibility and authority to some images, but in each case the viewer is persuaded by a unique set of circumstances. In other words, the ideology of perspectival images varies according to place and time. As Mirka Benes has demonstrated, Falda's perspectival views contributed to papal and aristocratic efforts to rebuild the power of the Counter-Reformation

1-9

Israël Silvestre, "Seconde Journée," *Les Plaisirs de l'Île Enchantée,* 1664.

Réunion des Musées Nationaux / Art Resource, New York

Church and aristocracy in seventeenth-century Rome.[42] Dal Re's panoramic prints show us how anxiety about status and the need for elite display during a time of social and political destabilization, the period of the Habsburg colonization of Lombardy, became a force in the shaping of visual and material culture.[43] Moreover, many eighteenth-century English estate views, whether painted or printed, served to naturalize the socio-economic consequences of the enclosure movement, the Black Act, and the Game Act. Those images asserted the status of property owners through the display of timber, game, and vast landed holdings set aside for non-productive purposes.[44] In a similar way, the bucolic and proto-Romantic views of Ermenonville included in an important guidebook to that property—the *Promenade ou*

itinéraire des jardins d'Ermenonville, 1788—portrayed the estate as common landscape unaffected by the social and economic tensions that in fact conditioned the French countryside [1-10].[45]

Views of property could also employ more subtle devices specific to image-making itself. For example, Jan Siberechts's large prospect, *Wollaton Hall and Park, Nottinghamshire* (1697), was painted in portrait format with the upper half of the image given over to sky [1-11]. In keeping with a device sometimes employed in Dutch landscape prints, the forms of the clouds in Siberechts's view mirror those seen on the ground, as if the lower realm were guided or affirmed by the upper. Similarly, a well-known depiction of Ermenonville included in Laborde's *Description des nouveaux jardins de la France* (1808–1815) puts a graphic convention to suggestive effect [1-12]. The

1-10
J. Mérigot, view of the House of Philemon and Baucis, Ermenonville, *Promenade ou itinéraire des jardins d'Ermenonville, 1788*; 1811.

Dumbarton Oaks, Research Library, Washington, DC

view overlooks a small lake in the Désert or Wilderness at Ermenonville, an extensive and ostensibly untouched natural area to the west of the château. The implied time of the image is at least twenty years earlier than the date of publication. The philosopher Jean-Jacques Rousseau, who spent his final years at Ermenonville and died there in 1778, appears in the left foreground gazing over the scene with his arms raised in a posture of astonishment or invocation. In the distance across the lake, a gap in the trees leads to a small structure, barely visible, that we know from other sources to be the so-called "cottage of Jean-Jacques Rousseau." Significantly, the visual axis is aligned perfectly with the vertical centerline of the image, a device pertaining to the logic of graphic composition. That arrangement helped anchor the image by making it appear balanced, but it also related the scene to older images of French gardens, in which a visual axis centered on a building divided the garden into bilaterally symmetrical fields. In that way, the graphic arrangement implied a historical transformation never suggested otherwise for this site, as if the Wilderness were a classical French garden overtaken by nature.

Finally, historic landscape views not only reflect the design of a landscape or the larger cultural context in which that work is situated. They also produce culture. Accordingly, they are highly significant historical agents as well as documents. To examine landscape views for accurate documentation of form with the aim of furthering an antiquarian project in garden history is a legitimate pursuit, and, in some cases, such examination may be fruitful, especially if the historian implements a multi-source approach that confirms the accuracy of the image. But to ignore the complexity of these images and to read them in reductive terms is not only to misuse them but also to commit a disservice to the historical legacy of the field of landscape architecture. To read historic landscape views exclusively as the documentation of garden form is to underestimate landscape history as a field of cultural study and, in turn, to suggest something similar about the contemporary field and profession. Although such images may not tell us much about specific form, they usually reveal a great deal about the broader significance of landscape in the formation of Western European cultural history. The aesthetics of gardens and the choices involved in their representation were and are related to elaborate patterns of status differentiation, among other concerns. Garden views were part of a complex discursive field related to social positioning and cultural authority, and they demonstrate the complicity of landscape in the workings of everyday life and in the shaping of culture.

1-11

Jan Siberechts, Wollaton Hall and Park, Nottingham, 1697.

Yale Center for British Art, Paul Mellon Collection, USA

1-12

Alexandre, comte de Laborde, view of the Wilderness at Ermenonville, with Jean-Jacques Rousseau, *Description des nouveaux jardins de la France*, 1808–1815.

Rare Book and Manuscript Library of the University of Illinois at Urbana-Champaign

1 See Dianne Harris and D. Fairchild Ruggles, "Landscape and Vision," in Dianne Harris and D. Fairchild Ruggles (eds), *Sites Unseen: Essays on Landscape and Vision*, Pittsburgh, PA: University of Pittsburgh Press, 2007, Chapter 1, and Martin Jay, *Downcast Eyes: The Denigration of Vision in Twentieth-Century French Thought*, Berkeley and Los Angeles, CA: University of California Press, 1993.

2 Ian Whyte has called this tendency to "read the history of landscape mainly through the history of landscape painting" a modernist shift among the subjective approaches to landscape studies. In a later passage, Whyte notes that "Paintings are now, however, considered to contain complex ideological messages, both superficially and at greater depth." Ian D. Whyte, *Landscape and History Since 1500*, London: Reaktion Books, 2002, pp. 19, 24. For a sample of opposing views presented in a single volume, see Linda Cabe Halpern, "The Uses of Paintings in Garden History," and Tom Williamson, "Garden History and Systematic Survey," both in John Dixon Hunt (ed.), *Garden History: Issues, Approaches, Methods*, Washington, DC: Dumbarton Oaks, 1992, pp. 190–191 and 63–64, respectively. For a more recent example in which paintings are assumed to be mimetically accurate, see Bridget Ann Henisch, "Private Pleasures: Painted Gardens on the Manuscript Page," in John Howe and Michael Wolfe (eds), *Inventing Medieval Landscapes: Senses of Place in Western Europe*, Gainesville, FL: University Press of Florida, 2002, pp. 150–170.

3 Michael Ann Holly, *Past Looking: Historical Imagination and the Rhetoric of the Image*, Ithaca, NY: Cornell University Press, 1996, p. xiii.

4 Lucia Nuti, "Mapping Places: Chorography and Vision in the Renaissance," in Denis Cosgrove (ed.), *Mappings*, London: Reaktion Books, 1999, p. 93.

5 John Harris, "Bird's-Eye Views at Yale," *Country Life*, 30 November 1978: 1822. See also, John Harris, *The Artist and the Country House: A History of Country House and Garden View Painting in Britain, 1540–1870*, London: Sotheby Parke Bernet, 1979, p. 41: "Not only is this a unique document in the evolution of the country house portrait, but it is also the only fully detailed view of the formal garden of a squire laid out before 1666."

6 See John Dixon Hunt, *Garden and Grove: The Italian Renaissance Garden in the English Imagination, 1600–1750*, Princeton, NJ: Princeton University Press, 1986, p. 131, and Harris, *The Artist and the Country House*, p. 41.

7 The authors gratefully acknowledge Mirka Benes and Mary Daniels for their assistance in obtaining the image by Aveline.

8 Kenneth Woodbridge called Liancourt "[t]he French garden most admired for its water in the early part of the seventeenth century." See Kenneth Woodbridge, *Princely Gardens: The Origins and Development of the French Formal Style*, New York: Rizzoli, 1986, p. 139.

9 See ibid., p. 138.

10 Dianne Harris, *The Nature of Authority: Villa Culture, Landscape, and Representation in Eighteenth-Century Lombardy*, University Park, PA: Pennsylvania State University Press, 2003, pp. 174–176. For more on Dal Re and his system of representation, see Chapter 1 of the same volume and Dianne Harris, "Landscape and Representation: The Printed View and Marc' Antonio Dal Re's Ville di delizie," in Mirka Benes and Dianne Harris (eds), *Villas and Gardens in Early Modern Italy and France*, New York: Cambridge University Press, 2001, pp. 178–205.

11 For an example that focuses on non-Western images, see James L. Wescoat, Jr., "Picturing an Early Mughal Garden," *Asian Art*, 2, 1989: 59–79.

12 See, for example, Mark Girouard, *Life in the French Country House*, New York: Knopf, 2000, pp. 11–29; Harris, *The Nature of Authority*, p. 62; Ralph Gibson and Martin Blinkhorn (eds), *Landownership and Power in Modern Europe*, London: HarperCollins Academic, 1991.

13 For a detailed discussion of these concepts, see Harris, *The Nature of Authority*, Chapter 3.

14 Christopher K. Currie, "Fishponds as Garden Features, c. 1550–1750," *Garden History*, 18(1), 1990: 23, 24, 25. See also Harris, *The Nature of Authority*, p. 128.

15 See, for example, Jorge Arditi, *A Genealogy of Manners: Transformations of Social Relations in France and England From the Fourteenth to the Eighteenth Century*, Chicago: University of Chicago Press, 1998; Jan Bremmer and Herman Roodenburg (eds), *A Cultural History of Gesture*, Ithaca, NY: Cornell University Press, 1992; Sylvana Tomaselli, "The Death and Rebirth

of Character in the Eighteenth Century," in Roy Porter (ed.), *Rewriting the Self: Histories from the Renaissance to the Present*, New York: Routledge, 1997, pp. 84–96.

16 For an example of estate views with figures delineated as recognizable individuals, see Richard Quaintance, "Unnamed Celebrities: Jacques Rigaud's Topographical Prints," *Cycnos*, 11(1), 1994: 93–132.

17 Louis Carrogis, known as Carmontelle, *Jardin de Monceau, près de Paris, appartenant à S. A. S. Monseigneur le duc de Chartres*, Paris: Delafosse, 1779.

18 View B from the later edition is reproduced in the exhibition catalog, *Grandes et Petites Heures du Parc Monceau*, Paris: Musée Cernuschi, 1981, p. 24.

19 See the plan of the Jardin de Monceau, dated 1788, showing the work of Thomas Blaikie, Bibliothèque d'Art et d'Archéologie, Paris, Fond Jacques Doucet, F8 I 103, reprinted in *Grandes et Petites Heures du Parc Monceau*, p. 43. See also David L. Hays, "'This is not a jardin anglais:' Carmontelle, the Jardin de Monceau, and Irregular Garden Design in Eighteenth-Century France," Chapter 11 in Benes and Harris, *Villas and Gardens*, pp. 323–326.

20 Carmontelle, *Jardin de Monceau*, p. 7.

21 See Hays, "'This is not a jardin anglais,'" pp. 304, 320–321, 323.

22 See Friedrich Melchior and Freiherr von Grimm (eds), *Correspondance littéraire, philosophique, et critique de Grimm et de Diderot*, 15 vols, reprint, Paris, 1879, XI: 372 (November 1776).

23 Ibid.

24 Raymond Williams, *Television: Technology and Cultural Form*, London: Fontana, 1974.

25 Malcolm Andrews, *Landscape and Western Art*, New York: Oxford University Press, 1999.

26 However, members of the seventeenth-century DeRossi family—the publishers who produced many of Giovanni Battista Falda's images of villas in Rome—considered Falda's work remarkable for its perspectival accuracy—another suggestion that some prints provide a degree of reliable documentary evidence of site conditions in their time. Francesca Consagra

has located documents among Giovanni Giacomo DeRossi's papers that clearly illustrate the desire for the implementation of a more correct use of perspective. DeRossi wrote that the key aspect of Falda's work was "views in perspective, delightfully done … better arranged and more correct than the others issued thus far." See Francesca Consagra, "The DeRossi Family Print Publishing Shop: A Study in the History of the Print Industry in Seventeenth-Century Rome," PhD dissertation, Johns Hopkins University, 1993, p. 418.

27 Although printmakers seem to have assumed they would profit from these ventures, they frequently did not do so. Documentary evidence indicates that both DeRossi and Dal Re died in debt and led lives marked by impecuniosity. On Dal Re, see Harris, *Nature of Authority*, pp. 21–22; See also Consagra, "The DeRossi Family Print Publishing Shop," p. 249.

28 See Erik Neil, "Architecture in Context: The Villas of Bagheria, Sicily," PhD dissertation, Harvard University, 1995; Erik Neil, "Emulation and Distinction: An Interpretation of Villa Building and Villa Culture in Seventeenth- and Eighteenth-Century Palermo," unpublished paper delivered at the conference "Architecture and the Baroque," 24 January 1998, Canadian Centre for Architecture; Tracy Ehrlich, *Landscape and Identity in Early Modern Rome: Villa Culture at Frascati in the Borghese Era*, New York: Cambridge University Press, 2002.

29 Chandra Mukerji, "Reading and Writing with Nature: A Materialist Approach to French Formal Gardens," in John Brewer and Roy Porter (eds), *Consumption and the World of Goods*, London: Routledge, 1993, pp. 445, 447.

30 See, for example, Mukerji, "Reading and Writing with Nature"; W. J. T. Mitchell, "Showing Seeing: A Critique of Visual Culture," *Journal of Visual Culture*, 1(2), 2002: 165–181; Norman Bryson, *Vision and Painting: The Logic of the Gaze*, New Haven, CT: Yale University Press, 1983; Hal Foster (ed.), *Vision and Visuality*, Seattle, WA: Bay Press, 1988; Jay, Downcast Eyes; Teresa Brennan and Martin Jay (eds), *Vision in Context: Historical and Contemporary Perspectives on Sight*, New York: Routledge, 1996; John Berger, *Ways of Seeing*, Harmondsworth: Penguin Books and BBC, 1972, p. 10. There are many others, but this selection offers a starting point. For the literature dealing with vision and landscape, see the special issue of *Journal of Garden History*, 14, 1994, guest-edited by D. Fairchild Ruggles and Elizabeth Kryder-Reid; D. Fairchild Ruggles, *Gardens, Landscape, and Vision in the Palaces of Islamic Spain*, University Park, PA: Pennsylvania State University Press, 2000; and Harris and Ruggles (eds), *Sites Unseen*.

31 For another mode of representing multiple aspects of landscape in a single view, see John Pinto's analysis of Piranesi's Lago Fucino print in which additional views overlap the plan as though pinned to the drawing. John Pinto, *Architects and Antiquity in Eighteenth-Century Rome*, Ann Arbor, MI: University of Michigan Press, forthcoming.

32 See the sources listed in note 30.

33 On the influence of the Galli di Bibiena family on artistic production in the eighteenth century, see A. Hyatt Mayor, *The Bibiena Family*, New York: H. Bittner, 1945.

34 Michael Baxandall, *Painting and Experience in Fifteenth-Century Italy*, Oxford: Clarendon Press, 1972, pp. 29–36.

35 See Woodbridge, *Princely Gardens*, pp. 154–155.

36 See Raoul de Broglie, "Le Théâtre de Chantilly," *Gazette des Beaux-Arts*, LVII, March 1961: 155–166, and Jean-Pierre Babelon, *Album du Comte de Nord: Recueil des plans des châteaux, parcs et jardins de Chantilly levé en 1784*, Saint-Rémy en l'Eau, France: Monelle Hayot, 2000, pp. 44–45.

37 Giovanni Battista Falda, *Li giardini di Roma con le loro piante, alzate e vedute in prospettiva*, Rome, 1680.

38 Giuseppe Zocchi, *Vedute delle ville e d'altri luoghi della Toscana*, 1744, and Gianfrancesco Costa, *Delle delicie del Fiume Brenta: Espresse ne'palazzi e casini situati sopra le sue sponde: Dalla sboccatura nella laguna di Venezia fino alla città de Padova*, Venice, 1750–1762.

39 See Harris, "Bird's-Eye Views at Yale," pp. 1820–1823, and Stephen Daniels, "Goodly Prospects: English Estate Portraiture, 1670–1730," in Nicholas Alfrey and Stephen Daniels (eds), *Mapping the Landscape: Essays on Art and Cartography*, Nottingham: University Art Gallery, Castle Museum, 1990, pp. 9–12.

40 See, for example, David L. Hays, "Figuring the Commonplace at Ermenonville," in Martin Calder (ed.), *Experiencing the Garden in the Eighteenth Century*, Bern, Switzerland: Peter Lang, 2006, p. 100.

41 Tom Williamson, "Garden History and Systematic Survey," pp. 59–78.

42 Mirka Benes, "Landowning and the Villa in the Social Geography of the Roman Territory: The Location and Landscapes of the Villa Pamphilij, 1645–70," in Alexander von Hoffman (ed.), *Form, Modernism, and History: Essays in Honor of Eduard F. Sekler*, Cambridge, MA: Harvard University Graduate School of Design, 1996, pp. 187–209.

43 Harris, *Nature of Authority*, passim.

44 On this topic, see, for example, John Barrell, *The Dark Side of the Landscape: The Rural Poor in English Painting, 1730–1840*, New York: Cambridge University Press, 1980; Ann Bermingham, *Landscape and Ideology: The English Rustic Tradition, 1740–1850*, Berkeley, CA: University of California Press, 1986; and Stephen Daniels, *Humphry Repton: Landscape Gardening and the Geography of Georgian England*, New Haven, CT: Yale University Press, 1999.

45 See Hays, "Figuring the Commonplace at Ermenonville."

2

Stephen Daniels

Scenic Transformation and Landscape Improvement: Temporalities in the Garden Designs of Humphry Repton

Time as well as space framed the landscape gardening of Humphry Repton (1752–1818), not least the passage of time which affected his designs on the ground. Shortly after Repton's death, it was reported that a number of his gardens were ruinous and overgrown. The biographical Preface to the 1840 edition of Repton's published works, probably authored by his son John Adey Repton, maintained that such a book would provide a more enduring record of his fame:

> Time makes unrelenting havoc with designs which, during the first ten or twenty years, may have afforded unmixed satisfaction. Young trees will ougrow their situations, while old ones will be uprooted by age or accident; flower-gardens which owed their charm to the light but fragile trellis ornament, or the constant culture of their elegant parterres, will fall into decay, or be neglected by their owners; while the facility with which any alterations may be made, aiding the love of change which is natural to most minds, in the course of years leaves no trace of that master-hand which had first laid the foundation of future improvement.[1]

It was a process that Repton himself recognized in his constant travels to commissions during the dramatic social and economic changes of his career, as changes in ownership and stewardship resulted in the abandonment or alteration of what he had proposed. As a landscape gardener who was not a contractor (like Capability Brown) but a consultant, preparing designs of meticulous, and sometimes fragile detail for others to implement and manage, his grounds for complaint may seem a little thin, but they effectively supported his insistence that his art of landscape gardening would endure more on paper, in writing and illustration, than on the ground.

The highly changeful, even revolutionary, period of British history in which Repton worked made him conscious of articulating time in his various works, in his designs but also his essays, verse, correspondence and autobiography. Especially in his later works, when he was conscious of his own personal and professional decline (correlating in his view with that of the country), fragments of the past and emblems of time passing figure prominently [2-1]. The ageing Repton played the part of the curmudgeonly English reactionary, in a theatrical view of his life and art, and this streak of showmanship reflected a consciously modern, forward-looking sensibility, and an enduring conviction, part pious, part playful, that the world might be marvelously, if only momentarily, transformed [2-2, 2-3]. In this chapter, I will chart various ideas and images of temporality in Repton's art. I will do so in the space between two processes, one sudden and spectacular, the other gradual and more mundane: scenic transformation and landscape improvement.

2-1
Tailpiece to Humphry Repton's memoir.
British Library, Manuscripts Department

Repton was already an accomplished landscape sketcher when he decided in his thirties to become a landscape gardener, and he refined his method in the picturesque style made popular by William Gilpin in his guidebooks. The poet William Mason told Gilpin:

> [he] can draw in your way very freely . . . by this means he alters places on Paper & makes them so picturesque, that fine folks think that all the Oaks &c he draws on Paper will grow exactly in the same shape and fashion in which he has delineated them, so they employ him & at great Price.

In characterizing Repton as something of a conman, for clients who apparently had little idea of how plants grew, Mason pointed out Repton's

2-2 *[below left]*
Frontispiece to Humphry Repton's memoir.
Private collection

2-3 *[below right]*
"Sunshine after Rain"
Humphry Repton, Fragments on the Theory and Practice of Landscape Gardening, *1816*

deployment of a Gilpinesque strategy as an advertising technique, improving the dull scene you saw with the delightful one you envisaged. Picturesque tourists used a variety of techniques to do this, compositional schema and optical instruments, and Repton deployed his own device, which became his trademark, a slide or flap overlain on a scene to show the present scene which, when removed, would show the landscape he proposed. The actual changes proposed might be relatively minor, say, removing a fence or trimming some trees, but the scenic effects were often striking, sometimes spectacular [2-4A, 2-4B].[2]

Repton attributed the idea of contrasting present and proposed scenes in landscape improvement to his friend, the Norfolk tree planter, William Marsham, although he was keen to claim the use of overlays as his own

2-4A, 2-4B *[opposite]*
"General View of Sheringham Bower," with and without overlay
Humphry Repton, Fragments on the Theory and Practice of Landscape Gardening, *1816.*

invention. The device of overlays made Repton's art suspect to some of his rivals in the profession of landscape improvement. It was an illusionistic sleight of hand, which made the present scene seem not only worse than it was but also a state of affairs which had somehow been papered over a landscape Repton was intent on uncovering, restoring to its proper condition. One critic reckoned it turned "rural improvement" into "rural pantomime." The serious business of landscape improvement, to be conducted with due regard to wider issues of countryside management, carefully planned as a long-term, sustainable investment (as we might now say) with due regard to soil and society, was to be turned into a form of frivolous showmanship, more akin to the illusionistic entertainments of London, the popular pantomimes, or plays given over largely to a succession of spectacular scenic transformations. We don't know enough in detail about the process of implementing the designs in Repton's Red Books to assess the degree to which more down-to-earth views of landscape improvement affected their theatricality. One powerful estate steward, the forester Thomas Davis, complained in his comments on the designs, that one of Repton's planting proposals at Longleat, felling old limes, planes, and elms near the house to make way for maples, thorns, and alders, was "a Stage trick" (Repton was pleased to report later that the trees were removed by a Spring blight, one of a number of instances where he noted Nature had overruled men in implementing his designs). Stewards, as Repton complained, were always altering his plans, many probably unconsciously in the process of making working drawings or staking out the ground, some no doubt advising their owners that certain features were neither practicable nor desirable. The Red Book, as Repton recognized, was valued for itself, as an object to be admired in the library rather than utilized in the estate office, an album of views, or as some saw it, a box of tricks.[3]

2-5A, 2-5B
Benjamin Pouncey after Thomas Hearne, an "undressed" park; a park "dressed in the modern style."

Richard Payne Knight, The Landscape: A Didactic Poem, *1794*

When Repton's antagonist, the picturesque connoisseur Richard Payne Knight, saw the Red Book for Tatton Park in a Pall Mall bookseller's shop window, as a way of raising subscriptions for Repton's first published volume, *Sketches and Hints on Landscape Gardening*, he used the device of contrasting scenes to parody professional landscape gardening in his didactic poem *The Landscape* [2-5A, 2-5B]. The first plate shows a pleasingly picturesque scene, of an Elizabethan mansion by the edge of an ancient forest, apparently untouched by improvement. The second shows the scene "dressed in the modern style," emparked, lawned, improved not much so in the style of Repton as of Capability Brown. It is a contrast which opens up some of the convoluted arguments over picturesque landscape. Brown was condemned by Knight for having no pictorial sensibility, Repton clearly displayed some, but not the connoisseurship, the knowledge of Old Masters, which Knight valued.

So while Knight commissioned Thomas Hearne (who had portrayed Downton, Knight's estate) to draw the views, he chose "the commonest English scenery" for the first view, "that I might not be supposed to take advantage of tricks of light and shadow of my own system"; and the engraver, Benjamin Pouncey had in the second plate "favoured that which what I condemn, by giving more breadth, in the light and shadow that there is, to the second plate rather than the first." Even allowing for this deliberate reduction of pictorial contrast (which Repton tended to overdraw), the argument is that the process of professional improvement is a dumbing down, as a taste for theatrical scenery reduced the landscapes of Old Masters.[4]

Repton might have taken the comparison of his art to pantomime more positively than his critics thought. He was a lifelong fan of theater of all kinds, pantomimes, masquerades, private theatricals, and the highly staged performances elsewhere in polite venues, from the pump room in Bath to the Lincolns Inn law courts. He liked dressing up, for one masquerade in the style of a Dutch burgermeister with a windmill in his hat, and, in a friar's habit for meetings of a Norwich fraternity, the College of United Friars, for whom he composed a gothic romance, *The Friar's Tale*. He wrote a play, *Odd Whims or Two at a Time,* which was staged by a touring company in East Anglia, in converted barns as well as purpose-built playhouses. He was a house critic for Boydell's Shakespeare Gallery, which marketed engravings of scenes from Shakespeare by modern artists. He could be high-minded about the illusionism of his art, drawing on Burke's aesthetic of deceptions and the performances at Drury Lane in Shakespearean roles of David Garrick and Sarah Siddons. But he was more of a popular entertainer and miniaturist. Walter Scott shrewdly and sympathetically compared Repton's designs to "a raree show omitting only the magnifying glass & substituting the red book for the box and strings." He was once observed on his way to a commission at Biggleswade Fair, enthralled by a puppet show, entertainments which often had highly elaborate scenographic effects, and the kind of themes, mythological, historical and topical, staged in large-scale London pantomimes. In his own fashion, Repton was managing the theatrical space between landscaping, tourism, painting, and gardening that had helped define polite society in the eighteenth century.[5]

There had been a well-established theatrical tradition of garden design in Britain since the time of Inigo Jones. Throughout the eighteenth century there were close exchanges between landscape art and scene painting, with many artists like George Lambert, Michael Angelo Rooker, and Philip de Loutherbourg working in the theater. Scenic tourism was consciously theatrical. William Gilpin's influential tour of the Wye is set out as an aesthetic program which has the visitors enjoying a switchback ride down the river, with shifting combinations of "sidescreens" (the banks) and "frontscreens" (the view towards a bend), a schema that not only suppressed information about the places passed through but also the claims of narrative, the images of progress that river scenes often encouraged.[6] London theaters were renowned for their scenography, with a repertoire of drops from pastoral and rustic scenes to sublime views of storms and volcanoes. Woods were stock scenes, cut to reveal prospects. Scene changing was rapid, with four or five scenes on each side of the stage thrust in or pulled back along grooved channels. Lighting and sound enhanced the effects. Successful scenery might draw more applause than the efforts of the actors. The mercurial de Loutherbourg was the most renowned designer. He made small models, with cut-outs, stained glass and candles, for scene painters and stage machinists to build on a large scale, and set up his own miniature theater for scenic performance. Spectacular transformations might be given a serious cultural significance, for example, de Loutherbourg's spectacular pantomime stage-craft and landscape art are linked to masonic rituals of physical and spiritual change, chemical theaters of transformation, a tradition strong on the Continent where it also found expression in garden design.[7] But for con-temporary English critics theatrical landscaping represented change that was superficial, sudden and temporary, a sequence with little sense of continuity or progress, all the more culpable for being part of the commercial culture of image and display, and its allure for capricious consumers, especially whimsical women.[8]

The anti-theatrical critique of landscape gardening is made clear in Jane Austen's *Mansfield Park* where the word "improvement" is corrupted by the specter of inordinate change, wanton alteration, and Repton is a brand name for money-minded delinquents: "Mr Repton... His terms are five guineas a day... Repton or any body of that sort ... any Mr Repton who would... give me as much beauty as he could for my money." Repton's projected visit to the locality is paralleled by that of the London scene painter for the private theatricals that disrupt the physical and moral fabric of the mansion, open-ing a dangerous space between theatricality and reality in which the novel's characters lose their bearings. Mr. Rushworth, the owner of Sotherton, an old-fashioned estate, is enthralled by Repton's transformations of a friend's place in a neighboring county "I never saw a place so altered in my life... I did not know where I was." There is a plan to similarly transform the par-sonage at Thornton Lacey, hatched over a card game called Speculation. It is suggested by the scheming Henry Crawford after he was also lost, when his horse lost a shoe on a hunting gallop, suddenly finding himself "in the midst of a retired little village between gently rising hills." Crawford plans to transform the parsonage and grounds into a conspicuous mansion and park, "the occasional residences of a man of fortune. . . a man of education, taste, modern manners, good connections" visible to "every creature travelling the road." The future incumbent, Edmund Betram, who has now recovered his moral bearings after losing them in the theatricals, resists, planning instead a careful program to conserve the landscape and its pastoral associations.[9]

And yet Repton subscribed to many of the conservative values expressed in *Mansfield Park* and a number of other tracts of the time. As early as his commission at Babworth in 1792, he was deploying contrasting scenes to criticize how "despotic fashion" would fell an old oak grove and flood a valley, wrecking the setting that fostered moral life, as portrayed in an old-fashioned conversation piece of the family playing music (with Repton joining in on the flute). The contrast was made more public, as a comment on the moral decline of the country, in the paired scenes entitled "Improvements" of 1816 [2-6A, 2-6B]. The first scene is as Repton sketched it when passing an aristocratic estate, a shady park landscape of old trees on one side, a wooded common on the other, with access between and a bench by the road for passing travelers The second scene is he saw it "improved" ten years later, so transformed that "I no longer knew or recollected the same place" until a passing laborer told him the story. The estate had been sold to an owner who had felled the old timber on the left and planted quick growing conifers in their place, enclosed and plowed up the common on the right, put up a paling to exclude access with a noticeboard announcing mantraps and spring guns.

The contrasting regimes are presented in contrasting optics, the gentle, lateral, landed circulation of the one contrasting with the rapid, forward, financial circulation of the other. It is an answer to a question Repton was frequently asked, "whether the Improvement of the Country in beauty has not kept pace with the increase in its wealth." Two points are worth emphasizing. First, the historical depth of the benign scene is documented by the testimony of poorer inhabitants. Second, it is scenery as well as society that has been sacrificed, the arena for the performance of duties, narrowed by a realism constructed through the imperative of money.[10]

Repton always endorsed the Burkean idea that landscape was something to be lived in, not just looked at. And the social pleasures of spectatorship were integral to country life itself, indeed, were important as ways of renovating venerable places, even as ways of sustaining a commitment to country life to counter-balance the lure of places of fashionable resort. Take an example. Around 1802, Repton was pleased to secure as a client a rich Liverpool merchant, Richard Walker, who purchased unseen an old mansion, Michel Grove, tucked away in the Sussex Downs. Walker and his wife (a key figure in the transaction, as were many women, to the dismay of Repton's critics) were keen both to restore signs of the estate's old pedigree (the church, the house, its manuscripts) and to introduce fashionable features like a prefabricated pavilion sent down from London for viewing the sea. The mansion itself seemed to have an exotic character, said to have been built by a Knight of Malta to imitate a Moorish palace he had seen in Spain, one appropriate to a region close to Brighton where Repton was to essay his

2-6A, 2-6B
"Improvements."

Humphry Repton, Fragments on the Theory and Practice of Landscape Gardening, *1816*

most theatrical house and grounds, the Royal Pavilion. Repton also fitted up the Walker's London mansion for a masquerade, with "flowery garlands and coloured lamps," the hottest ticket in town, attended by the Prince of Wales as well as Repton himself. Walker died soon after and Repton fondly recalled how Walker and his wife "burst upon the World of Fashion like meteors that cross the sky," and the happy days he spent with them in Sussex and London. Repton told another Sussex client, Sir Harry Fetherstonehaugh, how he enjoyed wandering around West End showrooms choosing drapes, mirrors, stained glass and argand lamps for his mansion at Uppark, "the effect will be magic." "On the verge of eternity," Repton found "the best recipe for happiness is to make the most of [such] trifles and I find more amusement in drawing a lamp or inventing a paperhanging than in designing a Palace or planning a Church."[11]

For a landscape gardener who valued his profession as an entrée to the best circles, and for whom the cultivation of friendship was an integral part of his improvement of scenery, the brilliant beau monde might burn out but made a more lasting impression than "the constant succession of new, and general acquaintances" forced on him as he traveled constantly in search of work. Repton made written sketches of clients in his memoir as a series of "portraits like those in a painter's room." They varied in "progress," "from the slight chalk sketch to the finished picture." This parallels the process in Repton's method of doing tinted drawings, as he set it out to instruct a client in 1812. In the style of the "progressive method" of drawing manuals and treatises, and with their socially progressive view of education and accomplishment, Repton sets out the stages of realizing a landscape for any beginner to follow.[12]

2-7
"An inquiry into the changes in architecture"

Humphry Repton, Designs for the Pavilion at Brighton, *1808*

Finally, I want to consider the antiquarian strain in Repton's style as it features in the idea of landscape gardening as a progressive force. Around the turn of the century, when he was still confident of his prospects and that of his art, Repton compiled a history of landscape gardening in which he anticipated "some great future change" following from knowledge of the scenery and buildings of India, as set out in the drawings of Thomas Daniell and William Hodges, and a commission to remodel the Royal Pavilion at Brighton in the style. He illustrated the history of styles from three periods of Gothic, through Grecian to Oriental (if the pavilion is rather hidden by rushes) [2-7]. Repton had always enjoyed Oriental styling for its contribution to the exotic, Arabian Nights, effect, but now assumed a more serious, academic interest in Indian scenery, based on Daniell's drawings of Indian architecture and also his son John Adey's antiquarian imagination. Indian architecture seemed both deeply ancient (with specimens discovered in archeological excavations) and highly modern, with scope for using new materials such as cast iron. The designs freely sample various building details from Daniell's *Oriental Scenery*, from Hindu temples to Mughal palaces. Indian scenery was also of course a trophy of British imperialism, rivaling the French appropriation of Egyptian culture, as Delhi was occupied in 1803 and the remains of the Mughal Empire secured.[13]

Events such as the collapse of the commission for the Royal Pavilion, and its passing to John Nash, from whom Repton had broken bitterly, taking his son John Adey into partnership, and the collapse of Repton's optimism about the prospects of his art and career, resulted in a revision of Repton's antiquarian views, or rather their location. All mention of Oriental scenery was abandoned,

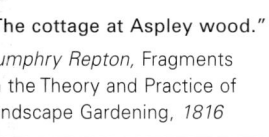

8

"The cottage at Aspley wood."

Humphry Repton, Fragments on the Theory and Practice of Landscape Gardening, *1816*

as he looked within English history for models, specifically Tudor history, as represented by buildings in his native East Anglia. This was the region in which Repton had lived and worked before he was a landscape gardener when he had taken a keen interest in its antiquities. It was a sensibility inherited by John Adey Repton, who trained as an architectural draughtsman in Norwich and became a Fellow of the Society of Antiquaries. John Adey developed a studious, pedantic antiquarian sensibility, making drawings of old buildings and details of old buildings, some of which he published in journals such as *Archaeologia* and *The Gentleman's Magazine* and an installment of John Britton's *Architectural Antiquities of Great Britain*. While an antiquarian sensibility was seen as an antidote to the disruptions, financial and political, of the period of the Napoleonic Wars, it was not a reactionary one. Indeed, John Britton valued it as a sign of cultural progress, one that was moreover made broadly accessible through the commercial press.[14] Inevitably, perhaps, there was a theatrical dimension, with the popularity of stock antiquarian scenery for gothic-style plays, although the more academic antiquarians were quick to condemn the historical gaffes and incongruities.

John Adey Repton's drawings are meticulously detailed, focusing mainly on detailed specimens. A few are, in his words, "restorations" (like the drawing of Oxnead Hall, Norfolk), researched through tracing out the foundations, collecting information from old inhabitants and consulting old books, including the published letters of the Paston family who lived at Oxnead. Here is the so-called "Elizabethan Gothic," a term the Reptons invented, which they valued in the manor houses of East Anglia and were keen to export to other regions, such as the Welsh borderland, which had no such architectural tradition. The style was activating and modernizing the Protestant imagination in much English antiquarianism, recognizing in this gothic a triumph over the barbarous ruins of a dark Catholic past. The Reptons were keen to build a gothic canopy for the tomb of one client's ancestor in Shropshire, a Protestant convert, "who died thro excess of joy on the accession of Queen Elizabeth to the throne" and was refused burial in church in the Catholic stronghold of Shrewsbury. John Adey incorporated drawings of details of old buildings into new designs with his father. The most striking was a lodge for the Duke of Bedford at Aspley [2-8] which incorporated no less than eighteen "fragments," remains and drawings of details of Tudor timber-framed buildings, from brick-nogging in King's Lynn to window tracery in Coventry and details from Tudor portraits, such as the maze and ornaments on posts and rails. Repton's last published volume, in which design is explained, is entitled *Fragments on the Theory and Practice of Landscape Gardening*, a title that both describes the break-up of his profession and the pieces that might be assembled in a spirit of progress to repair the divisions of the time.[15]

1 Humphry Repton, *The Landscape Gardening and Landscape Architecture of the Late Humphry Repton Esq*, ed. J.C. Loudon, London, Longman and Co.,1840, p. 3.

2 Stephen Daniels, *Humphry Repton: Landscape Gardening and the Geography of Georgian England*. New Haven, CT: Yale University Press, 1999, pp. 4–5.

3 Ibid., pp. 4, 150.

4 Ibid., pp. 112–113.

5 Ibid., pp. 4–6.

6 Stephen Daniels, S. Seymour, and C. Watkins, "Border Country: The Politics of the Picturesque in the Middle Wye Valley," in M. Rosenthal, C. Payne and S. Wilcox (eds), *Prospects for the Nation: Recent Essays in British Landscape, 1750–1880*, New Haven, CT: Yale University Press, 1999, pp. 157–182.

7 Daniels, *Repton*.

8 Stephen Daniels, "Gothic Gallantry: Humphry Repton, Lord Byron and the Sexual Politics of Landscape Gardening," in Michel Conan (ed.), *Bourgeois and Aristocratic Encounters in Garden Art, 1550–1850*, Washington, DC: Dumbarton Oaks, 2002.

9 Stephen Daniels and Denis Cosgrove, "Spectacle and Text: Landscape Metaphors in Cultural Geography," in J. Duncan and D. Ley (eds), *Place/Culture/Representation*, London: Routledge, 1993.

10 Daniels, *Repton*, pp. 12–13, 52–54.

11 Ibid., pp. 153–154.

12 Ibid., pp. 21–22; Humphry Repton, "A Few Hints Concerning Landscape Sketches," 1811, Getty Research Institute, 86-A248.

13 Daniels, *Repton*, pp. 191–205.

14 B. Lukacher, "Britton's Conquest: Creating an Antiquarian Nation," in the Frances Lehmen Loeb Center, Vassar College, *Landscapes of Retrospection: The Magoon Collection of British Drawings and Prints*, Poughkeepsie, Vassar College, 1999, pp. 1–40.

15 Daniels, *Repton*, pp. 18–20, 179–180.

3

Walter Hood

Color Fields

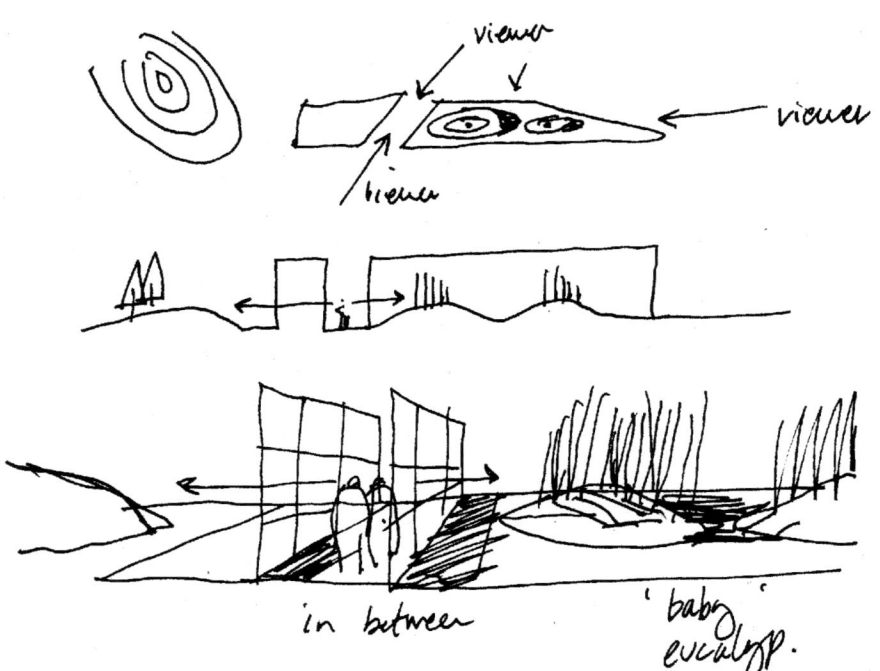

An act of creation, articulated even in the most simple terms, can be a momentous occasion in a designer's practice. A consciousness of the moment informs the abstract idea, a clarity embedded in the gesture drawing. On the famed paper napkin, on trace, or on a scratch pad, the sketch records the seed of an idea and the beginning of a design. Site and planning issues, program and environmental opportunities and constraints, initiate the struggle to create order from chaos—but in some respects, these functional criteria take a back seat. Whether in the plan view, perspective, section, or in combination with other drawing types, the gesture drawing attempts to describe both physical relationships and qualitative information. Formal ideas such as rhythm, balance, and repetition, describe order while dark lines, light lines, color, and tone focus the sketch.

The concept drawing may lead the designer to construct other two- and three-dimensional representations, furthering the process of design. But hopefully the initial gesture remains in the long process of the development that leads to built work. To progress from paper to space represents a great leap of faith. We can only hope that through the rigor of developmental drawing and modeling the initial gesture will, in the end, survive and permeate the work when realized [3-1].

3-1 *[opposite]*
Walter Hood. De Young Museum, San Francisco, California, 2001. Landscape concept sketches.

3-2
Walter Hood. Jackson Performing Arts Center, Jackson, Wyoming, 2002.

Hybrid drawings illuminating the genesis of the landscape concept.

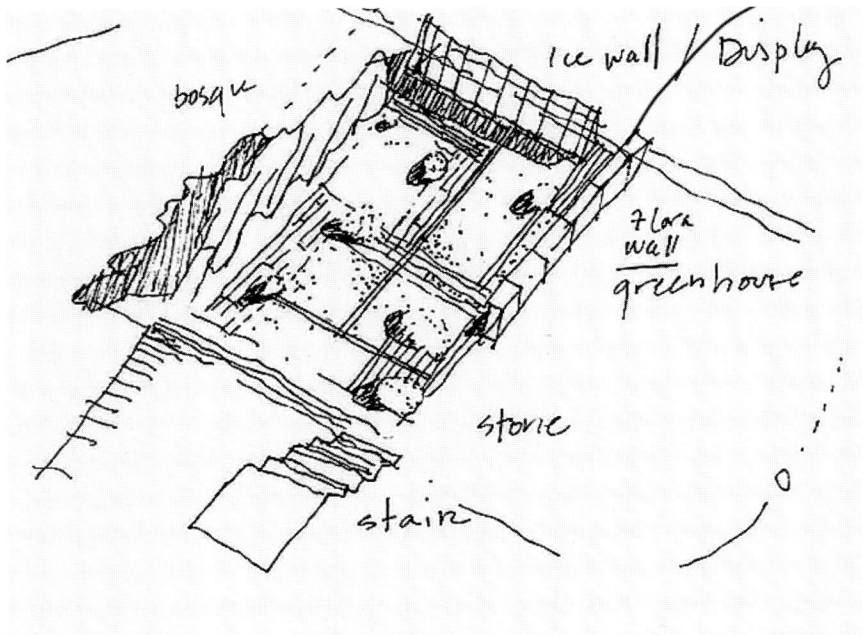

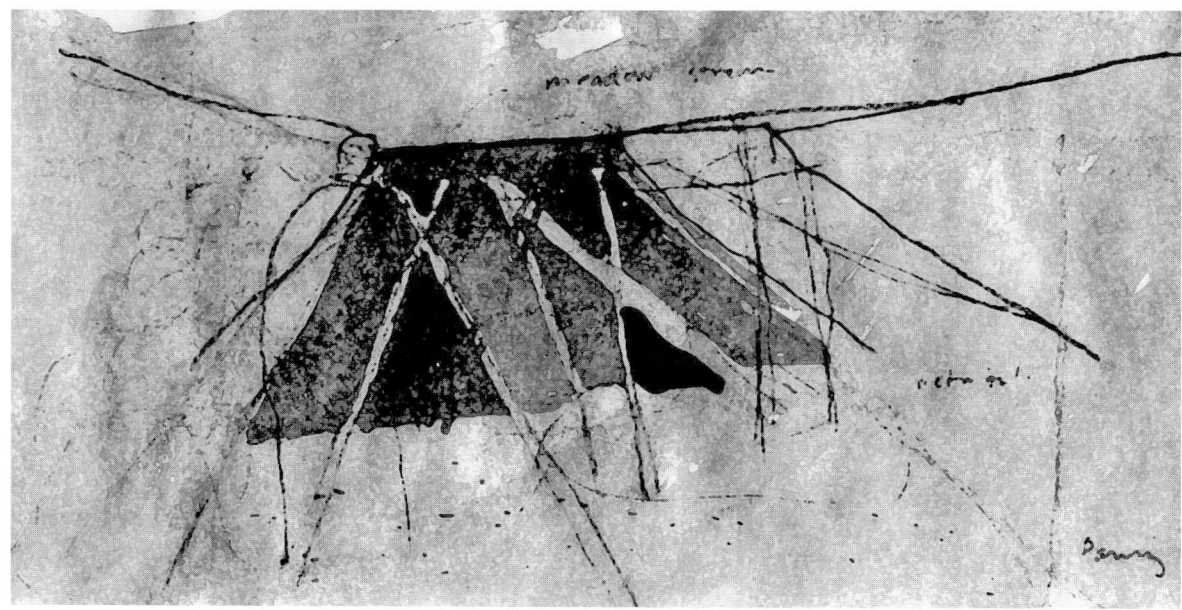

CONCEPT DRAWING #2

Concept drawings provide the framework for organizing our ideas. They are strategic representations that force the designer and viewer to discover additional possibilities. The designer utilizes conventional drawings in hybrid ways to transcend traditional operations in the hopes of leading us down a different path; they are experimental operations that stretch the imagination. By combining various drawing types—paraline projection, plan, section and perspective—new sets of relationships emerge on the page. The drawings become a sort of graphic text, a new notational system that organizes ideas about space, scale, and proportion [3-2, 3-3].

THE FIELD SKETCH

As a vital part of any landscape architect's education, the field sketch record works like others through careful observation and analysis. Whether capturing masterworks, analyzing the archetypal, or traveling to see at first hand the great buildings and landscapes, drawings developed from critical observations coerce us to better understand the subject. The field sketch can be quick and crude or laboriously detailed. Sitting and observing supports our study, with every line and detail itself a subject of study. Every glance and mark on the paper etch the subject viewed within our memory. The

3-3
Walter Hood. Autry National Center, Los Angeles, California, 2006.
Collage as gesture.

attention needed to accurately depict the subject offers the designer the opportunity to mentally reconstruct the subject before him or her. The sketches become part of the designer's personal catalog, awaiting future works that may reference or build upon the lessons learned through observation and drawing—whether these be compositional strategies, particular spatial relationships, or details.

THE PAINTER

Danger: landscape architects can become trapped and confined to the lines of the sketch they first put upon the page. At this point the drawing ceases to be the vessel for experience and observation in the environment, and becomes itself a subject of uncritical appreciation. To better understand the ideas possessed in the sketch, it can be useful to reinterpret what we have first drawn. One method is to reverse the relation of figure to field, by painting the negative space, by treating it as a positive [3-4].

The twentieth-century painter Phillip Guston utilized this reinterpretation of his early work, revisited their figuration through interpretive revisions in form and color. Certain figures remain in the later work, melding together with the space of the canvas. These fragments from prior paintings distill

3-4
Walter Hood. *Red Roma,*
water color and gouache on
paper, 1997.

gesture and form; they represented, to Guston, that which was most important. This is, perhaps, the closest example from painting of what the noted English architect Peter Smithson—a co-founder of the mid-century movement Team 10—characterized as the "purity of impulse": the immediacy of the designer's actions in the early stages of design. In this characterization, Smithson emphasized the extemporaneous act of drawing that sifts through the layers of information, conditions, and specific experiences of a project site, condensing them into a simple gesture on the page. It is a pure response that at least for a single moment stands free of objective criteria, as the scale and specificity of the subject give way to a clarity of approach.

Works by Henri Matisse also illustrate this process of distillation. His *View of Nôtre Dame* depicts the cathedral along the Seine as viewed from his studio. In subsequent years Matisse painted the same view several times, but in each new iteration the lines on the blue field allowed the painted space to consume, yet retain, the essentials of the earlier figurative paintings.

Matisse captured the essence of landscapes and condensed them into color fields. His representations trigger memories of particular places and objects and times, although they vary widely in their degree of abstraction. Matisse wove form, color, and objects into a limitless spatial field in which his conversations with personal experience continue. In his work we can feel Paris, Spain, and Africa all at once, as a whole, and the collective experience of these places exerts a presence even in paintings that picture other places. We can breathe the thick air of the Mediterranean and hear the Moors as Matisse, somehow, is able to combine the physical and metaphysical qualities in ways that transcend the eclectic and mundane.

One also recalls a passage by Quincy Troupe describing the jazz musician Miles Davis. Walking into a room where Davis was playing, Troupe thought one could hear the music of Africa, Brazil, Harlem, and Europe—all of it was there. One senses this compression of feeling and knowledge as well in works by Guston and Matisse.

EPILOGUE

Color fields do not replace the concept sketch, model, or other ways of recording, thinking, and making. They are a way to condense and remember, a way to work out the many fragments left from experiences in the world; most of all these paintings are a means to share with others—away from the constrictions of the making of landscape architecture [3-5].

3-5
Walter Hood. Piazza, Rome, Italy, 1996.

A color field. The interest lay in the repetition of landscape and built objects: fountain, church, and window. The color field defines the space they occupy.

4

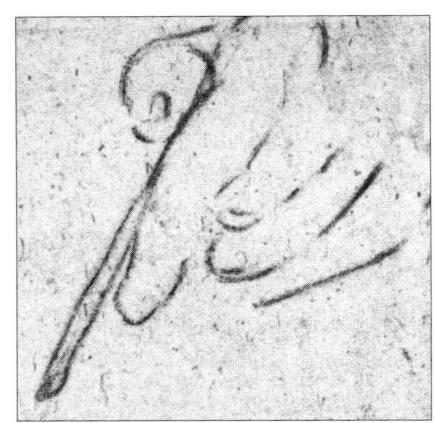

Chip Sullivan

Observation and the Analytical Representation of Space

"Drawing is the root of everything," proclaimed Vincent Van Gogh.[1] He reminds us that knowledge derives from close observation and analysis, and that drawing is the most direct method of recording one's observations. The very act of recording something increases one's perception of it, and heightened awareness is crucial to learning. Analytical drawing in particular is a process of observation that reveals a way of thinking. To analyze is to take something apart, figure out what it means, and put it back together. Analysis is concerned with training our eyes and hands to interpret the visual world from many points of view. We transfer information from reality onto paper through the link between eye and hand. The drawings we produce incorporate these layers of dissection and establish a framework for potential creative breakthroughs.

Leonardo's familiar portrait of his own hand in the act of drawing manifests the essential connection between the hand and eye [4-1]. His numerous sketchbooks clearly illustrate his thought processes and demonstrate his agile coordination of hand and eye in the simultaneous analysis of multiple problems or conditions. On a single page of his sketchbooks one can find both an analysis of a man's head as a three-dimensional form and—in the lower left-hand corner—a study for a domed castle [4-2]. A master of looking into the interior of life, Leonardo reasoned and expressed life's mysteries through drawing.

A critical journey that Filippo Brunelleschi made to Rome in the mid-fifteenth century clearly illustrates the relationship between analytical drawings and discovery. With a shovel, a ruler, and a pencil, Brunelleschi traveled to the Eternal City and began to excavate, measure, and analyze the ancient Roman ruins then being uncovered. By documenting the ruins of the Baths of Caracalla and envisioning how the fragmented remains had once fit together, he imagined and came to understand how the Romans constructed the large vaults that spanned the baths. Brunelleschi returned to Florence a changed man and revolutionized architecture with his construction of the dome for Santa Maria delle Fiore. From his sketches, we see that the unself-conscious documentation and representation of the world around us are fundamental to the expression of our visions and to our intuitive compre-hension of order. Analytical drawing is the graphic language we use to articulate the rational.

"Working drawings," or construction drawings, are documents that explain to others how a space or object is to be made. Analytical drawings, on the other hand, are "thinking drawings" in that they can explain—if at times only to ourselves—how a space is conceived; as such, they reveal the consciousness

4-1 [opposite above]
Leonardo da Vinci. *Codex Atlanticus*, folio 283. Sketch of left hand, c. fifteenth century.

Drawing connects the mind's eye with hand and paper.

© *Biblioteca Ambrosiana, Milan*

4-2 [opposite below]
Leonardo da Vinci. Study for St. James the Greater and a corner pavilion for a castle, c. 1495–1497.

"Thinking drawings" illustrate several ideas simultaneously.

The Royal Collection © 2006, Her Majesty Queen Elizabeth II

of a place. "Thinking drawings" illuminate the creative process behind an object or space and render the invisible visible by revealing relationships that may not be obvious to a viewer upon first inspection. They are learning tools that aid in the discovery of underlying spatial relationships, patterns, proportions, and systems.

Eric Sloane, a devoted student of the American vernacular landscape, used analytical pen and ink drawings to understand how the machines and tools of the past were fabricated and used. In a drawing of a late nineteenth-century grist mill, Sloane diagramed each individual part and reconstructed the machine through a progression of details and processes [4-3]. He began his visual explanation diagramming a section based on an anonymous mill-wright's sketch in which each of the working parts was labeled. The original drawing was then enlarged and expanded into a perspective section showing how the mill processed corn. Arrows illustrate the flow of the corn through the mechanism. The function and operation of the grist mill become clear as we visually move through the sequence of drawings that Sloane so carefully created, with a clarity unimaginable using words alone.

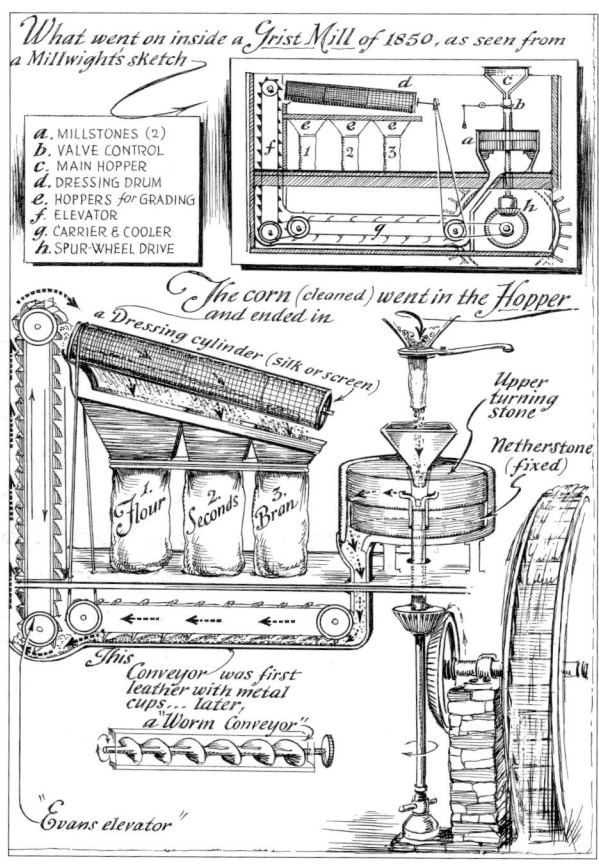

The satirist Rube Goldberg constructed elaborate sequential drawings to explain complicated processes of cause and effect [4-4]. His analytical drawings, as a means to a hypothetical end, investigated how things worked and interacted. "The unnecessary is the mother of invention," Goldberg once proclaimed. What is the most difficult way you can do something? How do you go about getting a gravy stain off of your coat? To answer this and other arcane questions, we navigate through his drawings and become keenly aware of the effects of time and motion. Goldberg was so successful at representing inefficiency and disorder that his name has become synonymous with any absurd and convoluted way of working. *Webster's New World Dictionary* defines the term "Rube Goldberg" as "any very complicated invention, machine, or scheme, laboriously contrived to perform a seemingly simple operation."[2] Through his step-by-step diagrams he makes the improbable real.

Serial drawings are effective tools for illustrating the complex relationship between time and space. Comic book artists are the innovative architects of time in that they are able to control sequence, timing, and movement through the medium of drawing. The comic page is a unique graphic form that frames a series of individual events that can be read independently

4-3
Eric Sloane. "Grist Mill of 1850," pen and ink drawing, c. 1955.

The graphic manifestation of historical research.

© *Courtesy of Mimi Sloane*

4-4
Rube Goldberg. "Hiding a gravy spot," c. 1920s.

Making the unbelievable believable.

© *United Features Syndicate Inc.*

CHEESE (A), AFTER STANDING AROUND FOR SEVERAL WEEKS, GROWS RESTLESS AND FALLS OF PLATFORM (B), HITTING SPRING (C) AND BOUNCING AGAINST ELECTRIC BUTTON (D) WHICH RELEASES ARROW HELD BY MECHANICAL CUPID (E) - ARROW CUTS STRING (F) DROPPING WEIGHT (G) INTO BUCKET OF WATER (H) - WATER SPLASHES ON TRAMP (I) WHO FAINTS FROM THE SHOCK AND DROPS AGAINST BOOMERANG-THROWING MACHINE (J) - BOOMERANG (K) SHOOTS ALL OVER THE PLACE AND FINALLY STRIKES END OF FOUNTAIN PEN (L) KNOCKING INK-BLOT ON PAPER- (M) - ERASER-HOUND (N) JUMPS AT PAPER TO RUB OUT INK-BLOT WITH HIS NOSE - STRING (O) SETS OFF AEROPLANE GUN (P) - BULLET (Q) HITS BOARD (R) WHICH SQUEEZES BULBS ON END OF DROPPERS (S) - DROPPERS CONTAIN INK THE SAME COLOR AS THE GRAVY SPOT - THE INK DROPS FALL ALL OVER VEST MAKING IT IMPOSSIBLE TO TELL WHICH ONE OF THE SPOTS IS GRAVY - AND YOU HAVE A FANCY VEST IN THE BARGAIN.

Simple method of hiding a gravy spot on your vest.
Tuesday, March 14, 1916.

and isolated in time, or reassembled into a complete narrative. The drawings can be read up close, far away, or just quickly scanned. They can be understood as having one overall message, or as capturing snapshots of individual moments. An early master of the medium was Frank King, who experimented wildly with unique formats of words and pictures in his comic strip *Gasoline Alley* [4-5]. In one full-page Sunday strip, King drew twelve individual panels that together created an aerial view of a beach scene; but, upon closer inspection the two main characters can be spied proceeding independently through each panel in sequential time. Each panel can read as a single image or the page can be comprehended as a chain of separate events. Our eyes process all the visual information at once and create a framework for understanding the story. Serial drawings are compatible with the way humans perceive space—initially as a totality, always seeking the gestalt or whole before the parts. The landscape is a continuous fabric that extends in all directions and is difficult to analyze until we establish a frame. Composite drawings and innovative graphic narratives help explain the dynamic and phenomenal qualities of the landscape.

The landscape architect Frank James used ideas of animated space and comic book sequencing to convey his own analytical design process. Using only a fine-tipped Rapidograph pen, James would create a gridded framework and with one continuous line develop numerous variations of a single design idea [4-6]. The line might continue to complete one image in several frames, or it might take off in a totally new direction, much like a stream-of-consciousness performance. His drawings grew even larger as one idea expanded into the next analytical sequence. James looked at the landscape design process as a chain of events imagined and expressed through drawing. He considered the entire design process as a cinematic script in which an idea becomes increasingly more defined as each individual image contributes to the totality. A sequence of static images can animate the landscape. Cinema is ultimately patterns of light and shadow on film, yet we believe the resultant pictures are real. The suspension of disbelief that we experience when individual frames are put into motion creates an imaginary visual world. "Thinking drawings" stimulate imagination and ideation, creating a reality with cinematic properties.

To produce analytical drawings we must first train our eyes to scan, filter, and select the essential characteristics of what we see. Looking itself becomes a vehicle for discovery. One must make assumptions and decisions in order to communicate his or her impressions or ideas. In organizing our visual experiences, we move from the general to the specific and isolate

4-5

Frank King. *Gasoline Alley*, c. 1929.

Individual events in time can be represented in a number of ways.

© Tribune Media Services, Inc. All Rights Reserved. Reprinted with permission.

significant visual elements. First impressions of an entire space can be quickly recorded with rapid gestural lines. More focused studies of individual parts will support the greater idea and clarify the design intention. Elements are interpreted and then reconstructed into meaningful compositions. Drawing provides a fresh vantage point to a traveler without a map. Even if we don't know exactly where we're going, by relying on intuition and drawing only what we see, a logic evolves to guide us onward. The drawing, in effect, becomes an analytical map.

Analytical drawings also help us understand the rationale behind a design and help define the meaning of a place. For instance, several years ago, while visiting the Villa Giustiniani for the first time, I sketched an image of the terrace wall encountered upon entry to the courtyard. The main garden appeared to be located directly above the wall [4-7]. However, after moving through the courtyard and ascending to the level of the balustraded terrace, I realized that the garden was actually separated from the terrace by a bridge. An axial pathway led from the bridge to a grotto at its far end, and a series of tall clipped hedges curved away from the path. A set of stairs, concealed by hedges, led to an upper level with additional parterres and a view of Rome in the distance [4-8]. From this vantage point, the axis of the parterres aligned with the main thoroughfare of the village, and in effect, the garden, the villa, and the town were visually connected to Rome. By drawing "through" the space and by recording the individual clues of each scene, the design logic was slowly revealed. Later, in the studio, the drawings became the basis for a new work that reflected my experience of the dramatically choreographed landscape [4-9]. Measured architectural drawings of the Villa Giustiniani—the conventional plan, section, and elevation—are dimensioned and rational but alone do not convey the dynamic experience of the space.[3] Alternative methods of graphic analysis are necessary to represent spatial relationships such as these.

Graphic representations not only analyze space but also clarify design concepts, and play a significant role in the design process. As a means of inspiration, any and all ideas can be recorded, developed, and evaluated. An image can be considered as a sum of individual parts, then examined on an enlarged scale. Using drawings to visually enter into and withdraw from space parallels the cinematographer's method of framing individual details and panning to move the viewer through a scene. In a similar way, one can use a variety of drawings to "think oneself through" a design to follow its development. Although an idea might reach a dead end, the designer can use the drawing itself to launch a new thought. In this case, by "thinking with a pencil" the individual parts cohere to produce a clear design idea.

4-6
Frank James. Detail of design process drawing, 1982.

A unique method of showing many variations on a single design theme.

Author's collection

4-7
Chip Sullivan. Pen and ink sketchbook drawing, 1984.

Recording observations of the Villa Giustiniani, Bassano Romano, Italy, seventeenth century.

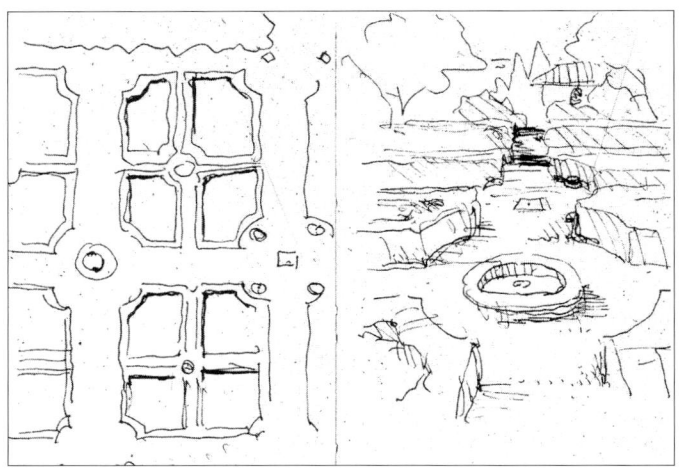

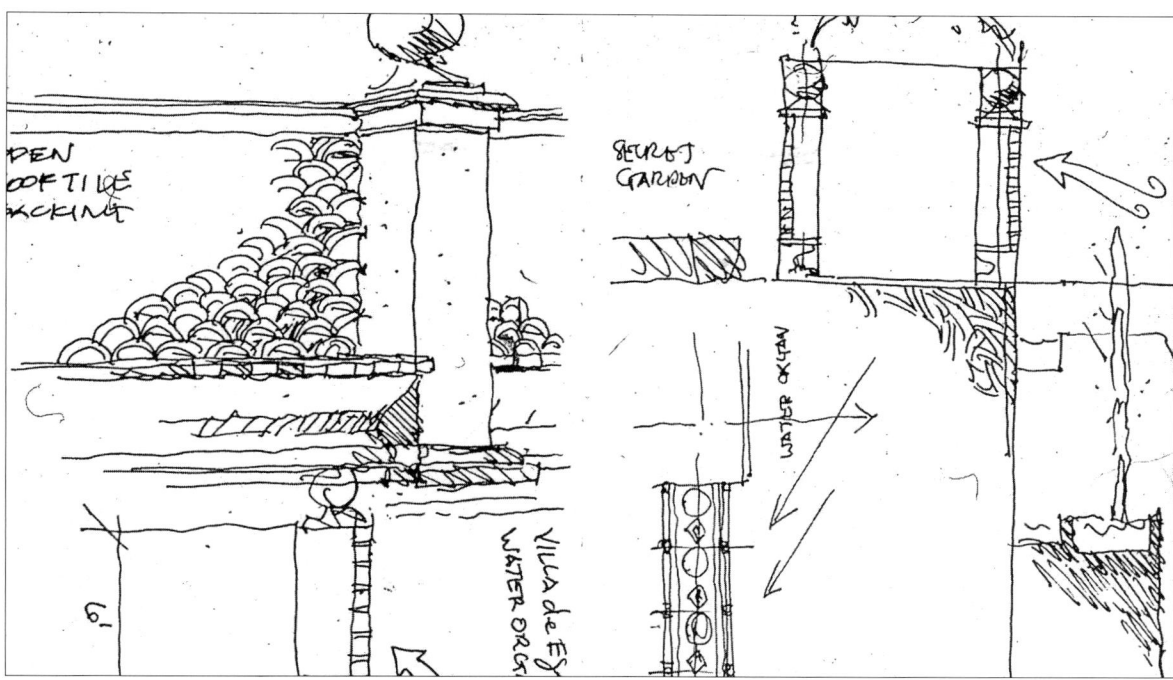

4-8 *[opposite above]*
Chip Sullivan. Villa Giustiniani,
pen and ink sketchbook
drawing, 1984.

Analysis of spatial complexities.

4-9 *[opposite below]*
Chip Sullivan. "Revelations,"
Villa Giustiniani, 1985.
Mixed mediums.

Site recordings inspired a box construction representing the memory
of hidden energy flows revealed.

4-10 *[above]*
Chip Sullivan. "The Cool Seat,"
pen and ink sketchbook
drawing, 1997.

Using drawings to decipher the
microclimatic qualities of the Villa
d'Este, Tivoli.

Indeed, drawing can be an adventure, like heading out on vacation with a pencil. Direct observation and analysis of one's surroundings may reveal hidden organizational systems and obscured meanings. One can try to envision a space through its designer's eyes to understand how and why the space was formed. A case in point: To decipher the particular climatic and passive design qualities of the Villa d'Este in Tivoli, Italy, sketches were produced to investigate the reasons why one corner of the garden felt especially comfortable despite the hot summer temperature. The air coming through the tiny openings in a decorative screen felt cool. Yet why was it cool? Further development of the drawing necessitated additional attention to the space and its details, and revealed a fountain beneath the screen. The air moving through the small apertures had, in fact, been cooled by the water [4-10]. Analyzing the space at both the macro and micro scales uncovered connections between seemingly disparate design elements. Detail investigations made more lucid the relationship of the parts to the whole, and described the climatic performance of the design.

In a sense, the landscape is endless; our observations are selective in that we focus on certain elements to the exclusion of others. Designers have considerable control over how people perceive and move through the landscape. To fully understand the effect of our landscape interventions on human activity and perception, it is imperative to observe and illustrate how the eye and the body can move, pause, and proceed through space. The eye can be fooled and nature manipulated by certain devices and interventions. Analytical drawings are useful in projecting how perspective may be distorted, scenery "borrowed," and romantic views manufactured [4-11].

4-11

Chip Sullivan. "Civic Center View Chamber," temporary, site-specific sculpture installation, San Francisco, 1995.

Elevational studies show reflected images and possible view lines.

In summary, like the game of billiards, design involves the action and reaction of different maneuvers. There is a consequence to every action and one must anticipate these possible, but invisible, trajectories through space. Watching and looking, we analyze and plan. A good player, like a good designer, employs a conscious strategy, while an amateur tries only to pocket one ball at a time.

Despite the abundance of digital technology and the popularity of electronic media, it is encouraging to believe that drawing will never become obsolete. The quickest, simplest, and cheapest way to launch a creative journey is to begin by drawing. "Art isn't a method for execution or a means of communication as much as it's a way of thinking," writes cartoonist Chris Ware. "And when you're drawing or doing whatever you do, you can't think it out in advance, and if you do, you kill it to begin with."[4] Ware confirms what Van Gogh and Goldberg concluded, that drawing is the mother of invention.

NOTES

1 Costa Ives, Susan Alyson Stein, Sjraar van Heugten, and Marije Vellekoop, *Vincent Van Gogh: The Drawings*, New Haven, CT: Yale University Press, 2005, p. 18.

2 David B. Guralink (editor-in-chief), *Webster's New World Dictionary of the American Language*, 2nd college edn, Cleveland, OH: William Collins and World Publishing Co Inc., 1976, p. 1243.

3 The Italian Renaissance estate at Bassano Romano, in the Roman Campagna.

4 Gary Groth and Chris Ware, "Understanding (Chris Ware's) Comics," *Comics Journal*, no.200, December 1997: 119–178.

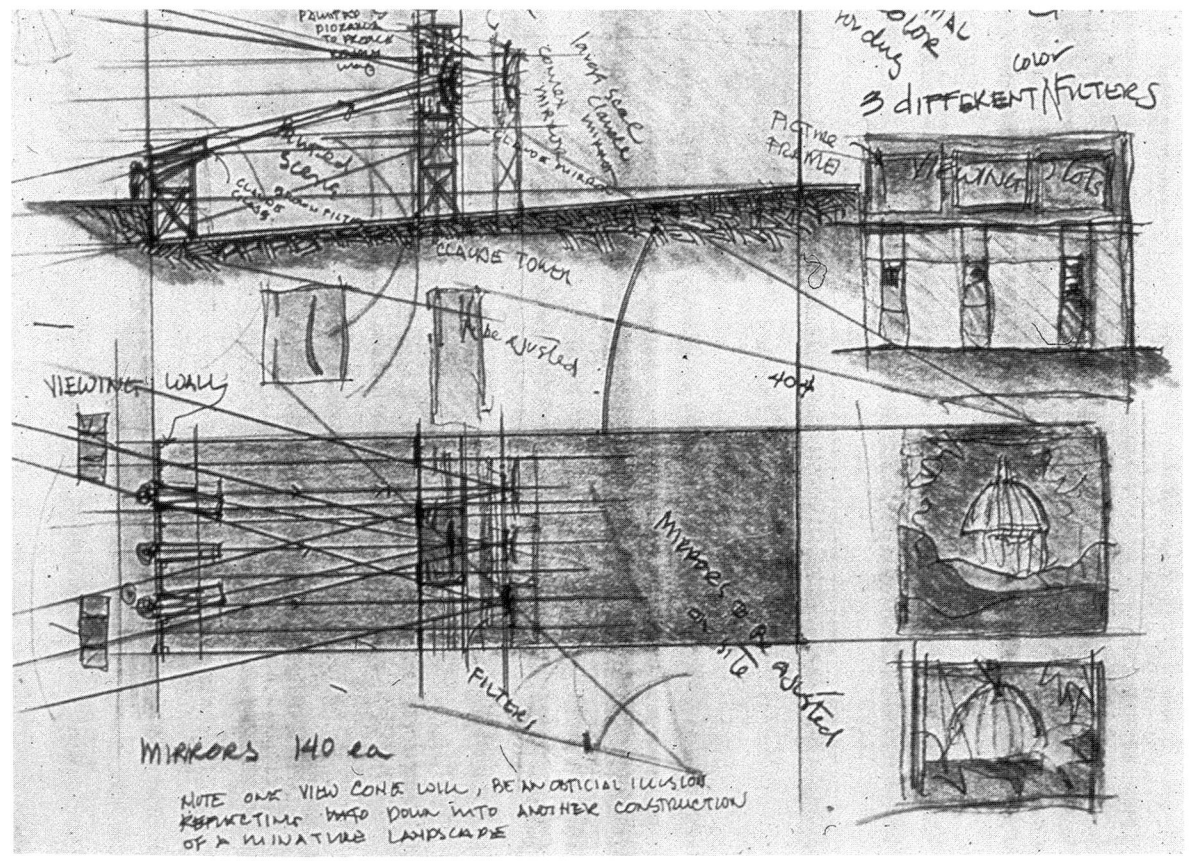

5

Thorbjörn Andersson

From Paper to Park

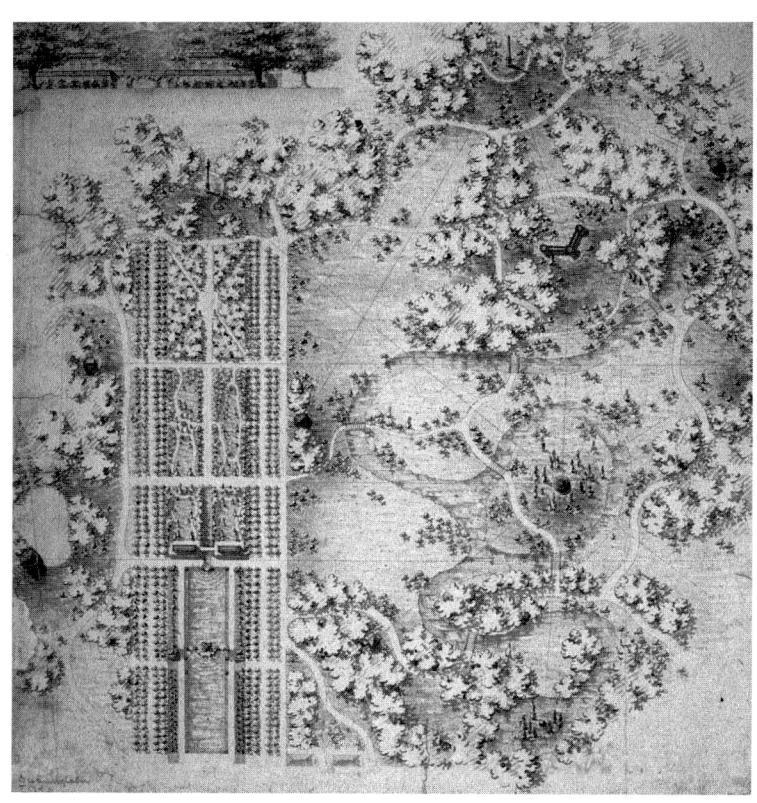

Few professions match architecture and landscape architecture in their need for graphic vehicles with which to mediate their ideas. In practice, they require representations that are at once abstract and simplified, yet legible and communicative; this is one characteristic that distinguishes the design professions from other artistic or engineering fields. The mastery of designing using pictorial or scaled representation adds immeasurably to the landscape architect's professional success [5-1]. Or, to put it another way, although talented in solving spatial and formal problems, a landscape architect does not function well professionally if he or she fails to develop graphic models that communicate those ideas precisely and persuasively.

However, design communication must be seen against another background. The relationship between the client and the landscape architect strongly influences the representational vehicles that are used—and needed—to realize any project. For example, a design for one's own kitchen garden may require neither drawing nor model because the client and the architect are invested in the same person. An idea and a spade are a sufficient means of transforming the abstraction (the idea for a kitchen garden) into an environment (the reality of a kitchen garden). If the project is more elaborate, make a model, or even mock up the design by staking out the terrain. However, the reason for this procedure is to ensure the quality of one's own idea, or to examine the means by which the design idea becomes a built reality.

5-1
Fredrik Magnus Piper.
Drottningholm Castle gardens,
Sweden, 1781. Site plan.
Watercolor and ink.
Royal Swedish Art Academy,
Stockholm

But even a kitchen garden may be a complicated project, requiring two separate minds to realize: the mind of the client and the mind of the gardener and/or the landscape architect. Realization requires clear communication, and the need for effective representations increases. The gardener Jean de La Quintinie planned and managed the extensive kitchen garden—*le jardin potager*—at Versailles, built during the 1660s. La Quintinie used neither drawings nor models. Real, colorful, fragrant and tasty fruits and vegetables presented his ideas. Large and luscious pears placed on the king's table in the winter month of February represented the gifts of the kitchen garden.[1] His achievement supported his request from the king for a free hand in establishing the royal kitchen garden, which was a minor miracle for the time. Constructed on wet land ill-suited for the purpose, the garden used innovative glazing techniques to create a type of open greenhouse, and radical espalier techniques for pruning fruit trees. To some degree these were techniques used in market-oriented gardening but on a scale undreamt of by commercial gardeners. To realize these structures, La Quintinie no doubt prepared drawings, but his principal mode of communication was fruits and vegetables—and his method successfully convinced his client, Louis XIV. In late eighteenth-century Sweden, Fredrik Magnus Piper used similar means to present his ideas to King Gustav III.

The communication process and the consequent need for sophisticated means of representation become more complicated in democratic societies in which a larger group is to be included in the decision-making process. Gardens in seventeenth-century France—as well as most classical landscapes until the first public gardens in the late nineteenth century—were created under political conditions in which very few people were in command of the decisions.[2] Paradoxically, a period of high quality in municipal park design—despite a quite limited use of representation by its designers—character-ized Stockholm's civic efforts during the 1940s and 1950s. Park production during this period became even more democratic, with a significantly widened user group, causing the design process to follow a convoluted road. The need for an effective means of communication, especially in the shaping of public space, became more critical as the number of the participants in the process increased.[3] For example, the downfall of the department was triggered by an incident in 1971. In that year, the remodeling of one of the city's most beloved parks, Kungsträdgården, involved the cutting down of a grove of elm trees. The citizens' protest against this action was close to a riot, and the Director of Parks resigned just a few years later. The elm trees are still there, however. Plans for the renovation came to nothing. In this case, the communication between the client (the citizens) and the landscape architect (the Director of Parks) had failed.

As a general rule, the need for effective graphic representation increases with the degree of complexity of the project. Factors influencing the type and number of images include:

1. The number of participants involved in the decision making. The element of more participants in the design process demands more elaborate represen-tation in order to develop understanding and generate reactions to the proposal. If the general experience of reading drawings is limited, that situation is further underlined.

2. The amount of detail in the program. If not clearly stated in the client's program, considerations must be sorted out by the designer in the design phase. A detailed program thus strengthens the client's position and in many instances makes the designer's task more focused. A strong client with clear vision, combined with legible programmatic issues, reduces the demands for conceptual representation.

3. The level of trust in the designer. A high level of trust in the designer can reduce the need for images, at least at the conceptual stage. Needless to say, this trust involves risk on the part of the client because the designer is

offered a larger mandate. The responsibility for the quality of the project, however, to a large extent becomes the designer's.

4. The level of innovation in the project. Repetition of known solutions and loyalty to established design ideas require less detailed drawings. In contrast, innovation in form, concept, material and function needs to be more thoroughly represented when taking more daring steps.

5. The nature of the project's construction. Today, the designer is commonly selected by comparison with other consultants. The criteria for such an evaluation often include price, competence, references, prior experience—and a preliminary sketch that proposes a design idea. In order to convince the potential client, the image in such a situation must be engaging and persuasive.

This chapter will examine three themes in the field of representation: (1) the terrain as vehicle; (2) the site plan; and (3) the perspective. Case studies drawn from Scandinavian landscape architecture will illustrate various interrelations between the client, the designer, and the program, and also show how the site and its conditions can be used almost as a program itself. The examples will display different representational balances between abstraction and realism, and between artistic expression and communication. The five factors cited above will provide the structure for examining these landscapes.

TERRAIN AS VEHICLE

A cartoon from the 1950s shows the Director of Parks in Stockholm, Holger Blom (1906–1996), pointing to a spot where he is instructing park workers to plant a large oak tree [5-2]. Under Blom's leadership in the 1940s and 1950s the Stockholm Park Department created so many parks—and with such consistent quality—that the landscapes they produced were regarded as the Stockholm School of Park Design. Its designers explored and refined the explicit use of natural landscapes as the basis for a new garden or a park, abandoning references to the picturesque or formal idioms. The rich landscape around Lake Mälaren and the capital city offered sufficient variation and usable elements to create prototypes for a new landscape types.[4]

5-2
Holger Blom, the Director of Parks in Stockholm, working on site. Sketch by Birger Lundquist, Stockholm Park Program, 1956.

Working with the natural landscape as a basis for the new parks required both strong discipline in the construction phase as well as an eye sensitive to the qualities of the place during the design phase. It also required a way of working where "adaptation" was the key word. Drawings were made only to show the rough concepts. The project was then staked out in situ, with adjustments and corrections made on site and in accordance with the ideas behind the proposal. This direct method shortened the gap between the representation and the realization.

The park department also utilized a technique by which the master plan and the details were immaculately developed. On the other hand, the "middle scale" between those two phases was rarely determined on paper. Instead, these design decisions were preferably made on site, using existing conditions as the skeleton for the new park. The general concept for the design was represented in the master plan, the middle scale was created on site, while the detailing was precisely determined on the drawing board: features such as hand rails, wood carvings, and pavement patterns.

Such a manner of working assumed that the parks would be constructed using in-house labor so that Holger Blom could control the execution as well as design phases. Over the years, sympathetic thinking developed between the designers and the construction teams, further reducing the need for drawings. The Parks Department worked only with its own designers, integrating the client–designer relation. The park program, a comprehensive document Blom formulated, laid the ideological foundation for the parks and as such provided verbal instructions that gave coherence to the projects even before they were initiated. With this program Blom created a framework within which his designers operated. That the designers, client, and construction teams shared the same budget also meant that Blom could order alterations and adjustment after construction had begun—until he achieved the intended the result. This economic flexibility allowed a graphic flexibility as well. Rather than requiring a complete set of detailed drawings necessary for competitive bidding, Stockholm's vertically-integrated system allowed minimum documentation at the start of construction and greater faith in the initial concept and the designers who created them.

Fredrik Magnus Piper (1746–1824) was one of the first professionally trained landscape gardeners in Sweden. Of noble birth, Piper shared the same birth year as Gustav III, with whom he had direct ties until the untimely assassination of the king in 1792. In Sweden, Piper is considered the country's foremost designer in the English landscape garden style and is the landscape architect behind the Haga pleasure grounds north of Stockholm, planned as the summer palace of the king. Abroad, Piper is known primarily for his renderings of the gardens of Stowe, Painshill, and Stourhead that have served as documents for subsequent study and restoration.

Piper studied mathematics and hydrology at Uppsala University from 1764 to 1766. Thereafter followed an education in engineering at the naval dockyards in Karlskrona. After further artistic studies at the Academy of Fine Arts in Stockholm, in 1772, he became the Superintendent of Public Works. Piper was an Anglophile, especially enthusiastic about English architecture and landscape gardening and in 1773 began work for the well-known British architect William Chambers.[5] Piper left Chambers' office in 1774 to continue his studies in France and Italy, but in 1779 he returned to England for two additional years.

It was during this second visit that Piper recorded and rendered the site plans for a number of British private estates, among others, Henry Hoare's Stourhead. These records are striking in their quality and degree of detail, in contrast to almost all other English garden plans.[6] The plan of Stourhead is accurately delineated with sight lines indicating the important views upon which the entire garden was conceptualized. Thus, the drawing not only presents but also analyses the park, depicting the relations between the principal areas of the park and showing them as the visitor would experience them at ground level. The brightness of Piper's watercolors reflects the atmosphere of the park with its glistening lake and rolling pastures.[7] A legend identifies each building. Framed picture boxes present views of the wooden timber bridge and the Palladian house. Piper's rendering of the terrain and his use of sight lines reveal his understanding of the sequential planning of the garden, but also testify to his study of drawings for fortifications. The curriculum at his former school, the Academy of Fine Arts in Stockholm, was based upon Parisian formulas that included fortification design and drawing as part of his studies in civil engineering.

Piper's highly original and compelling draughtsmanship earned him a position in English garden art as well as that of Sweden. England's formal gardens and hunting grounds are well documented—understandable considering the country's long tradition of topographical recording. However, the precision and depth of Piper's drawings have few equals. The famous *Britannia Illustrata* contains eighty bird's-eye views, all engravings. Piper, though, worked with pencil, ink, wash, and watercolor. Noteworthy is his way of showing relief through shading, a technique also evident in his plans of Stowe and Painshill. This technique became one of his trademarks after his return to Sweden in 1780.

Upon his return to Sweden, in 1780, Piper was appointed court surveyor and the crown entertained high hopes for him. The king was then in the midst of an intense period of work on the English landscape garden at his

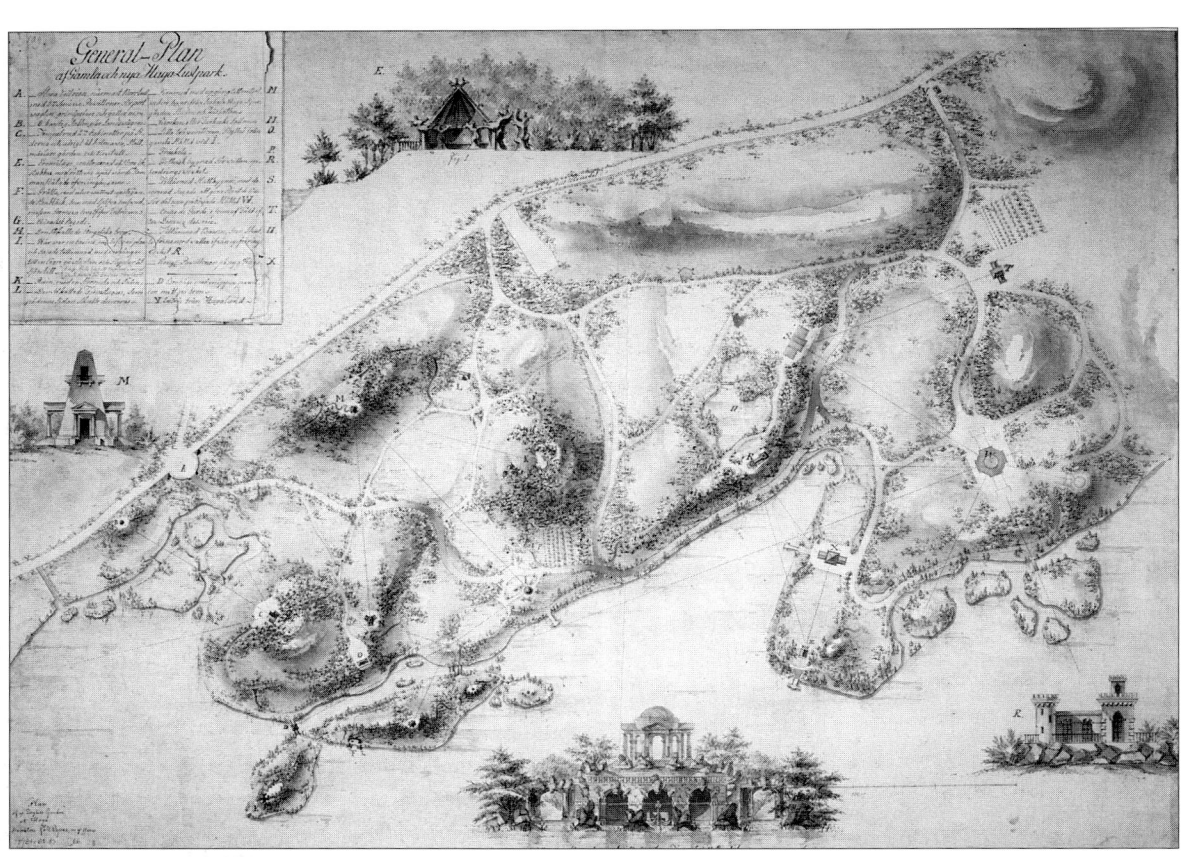

castle in Drottningholm and had developed his own proposal for the property [see 5-1].[8] Piper extensively revised the king's plan, in which all regular
features were removed while greater contrast between dense groves and
open meadows was added. Probably taken as criticism by the king, who
dismissed Piper from the project, it signalled a crack in the trust between
client and designer. This could be the reason why Piper, although at the time
the most qualified professional in the country, built relatively little. Piper
was probably also a victim of the times. Noble amateurs were interested in
making gardens during these years—for example, the banker Henry Hoare
at Stourhead, and King Gustav III at Drottningholm—and Piper's bitterness
about the lack of respect for true professionalism is apparent in the manuscripts he left behind.[9]

The master plan by Piper for Drottningholm was made in 1781. It shows the
existing formal garden, designed by Nicolas Tessin the Younger in 1681, and
the English landscape garden revised by Piper to the west; the two parts are
joined along a knife-sharp edge, neither one interfering or even acknowl-
edging the presence of the other. The sight lines typical of a Piper plan also
appear in this site drawing—the only gesture toward weaving the two parts
together: the *point de vue* in the French part connects in a diagonal view with
the central motif of the English park, a structure that crowns the Monument
Isle. Piper's plan renders the vegetation with individual characteristics, a
rare practice during the earlier period of formal design. In the Drottningholm
site plan, broad-leafed trees contrast with conifers and species used to form
allées differ from those that comprise groves and clumps.

5-3
Fredrik Magnus Piper.
Brahelund (Nya Haga),
Sweden, 1787. Sketch site plan.
Watercolor, wash and pencil.

*Royal Swedish Art Academy,
Stockholm*

Despite Piper's lack of diplomacy, the king gave him a renewed commission
for the new summer palace at Haga [5-3]. Relations remained frosty, however; Piper designed both the site plan and made sketches for the building
program, but most of the follies, temples, and buildings were produced by
other designers.[10] To Piper must still be attributed the overall design as well
as the positioning of the follies and pavilions. These structures marked the
highest points in the terrain and were set at suitable distances from each
other that coincided with important vistas and intersections. Typical of the
English landscape manner, all architectural features such as terraces, parterres,
or courtyards were banished and the great lawn runs directly up to the
façades of the buildings. The Haga plan shows Piper's mastery of the shading
technique through which the hilly terrain is emphasized with darker tones and
the undulating shorelines are underlined with strokes of dark blue watercolor.

Piper's beautiful finished wash and watercolor studies of English and Swedish
gardens suggest he had publication in mind while preparing them. Perhaps

this derived from the frustration he must have felt by having so many of his designs unimplemented. The desired publication never came to realization during his lifetime, however.[11]

Between Fredrik Magnus Piper and the Stockholm School is some 200 years. Despite this gap of two centuries, two direct parallels exist nonetheless. The first is obvious: both use the assets of the regional landscape as their point of departure. The second is less evident: for both, instructions for developing the site were combined with a master plan that directed the planning of the whole. Thus, each acknowledged both the abstraction of the overall plan and the reality of human experience on site.

THE SITE PLAN

Of all the landscape architect's drawn views, the site plan is the most ubiquitous and the most influential. From it derives the structuring of the site, reinforced by further study through the section and planting plan. Erik Glemme (1905–1959) was active as the head of the architectural department at the Stockholm Parks Department under the leadership of Holger Blom, discussed above. Glemme's site plans often included a level of detail and artistic skill that transmitted a clear impression of what the project was to be. They thus went far beyond being only instructions to the construction team.

The site plan of the cliff gardens in the Vasa Park, designed in 1949, illustrates this ambition [5-4]. The plan is drawn in ink using lines without rendering, yet still provides a clear sense of the intended atmosphere. The scale is 1:50, a sufficiently large enough scale to indicate details and materials as well as give a sense of space. The square stone wall surrounds a rich paving pattern of stones and pebbles, laid out in a manner that recalls the mosaic pavements of Moorish gardens such as the Alhambra, which Glemme mentioned as a source of inspiration. Vertical elements such as small fountains, flower beds, and a single tree interrupt the continuity of the pavement. Every perennial flower is depicted. The tree is drawn in a naturalistic manner and relates to several larger trees standing beyond the walled square, thus creating a link between inside and out. The lively brook located outside the walls produces a similar effect, bouncing playfully between little pools that descend the steep hillside, contrasting with the more tranquil pools within the stone enclosure.

Glemme's site plan for the cliff gardens is not a construction drawing. It contains no measurements nor can it be perceived as an assembly of details, as working drawings often are. Instead, it is more a narrative of a designed

5-4

Erik Glemme. Vasaparken, Stockholm, Sweden, 1947. Site plan. Ink on tracing paper.

Stockholm City Parks Department

environment represented in a naturalistic way. The drawing lacks abstraction; all forms, spatial circumstances, and materials are depicted as close to their actual appearance as possible. Yet the drawing is sufficiently accurate to be measured directly. The precision of the irregular joints in the walls and the specific shapes of the paving stones suggest that this plan could have been used at the construction site. Considering the integrated design construction practices at the Stockholm Parks Department, this might well have been so. The only additional material for the cliff gardens in the Vasa Park now found in the departmental archives is a series of small-scale sections and one single technical drawing showing a typical part of the pavement. Perhaps Glemme's pictorial plan served far more than a pictorial purpose.

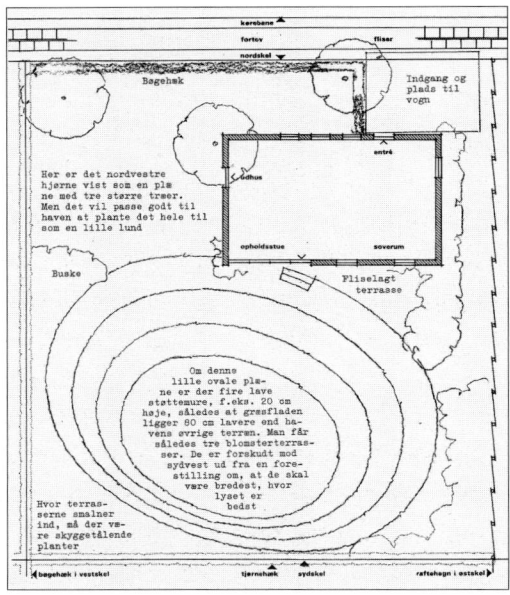

The Danish landscape architect Carl-Theodor Sørensen (1899–1979) began his own practice in the 1920s. For a seven-year period he also served as Professor of Landscape Architecture at the Royal Academy of Fine Arts in Copenhagen, and greatly influenced its curriculum. Sørensen was also a prolific writer with numerous articles and a dozen of books to his credit. Two of these are of special interest: the comprehensive study of European garden art, *Europas Havekunst,* of 1959 is one of them. In this book, Sørensen traces the stylistic origins of the high styles from the regional landscapes in which they are found. In *Utypiske haver til et typehus: 39 haveplaner* (Untypical Gardens for a Typical House: 39 Garden Plans, 1966), he produces a series of garden ideas all presented in a similar way [5-5]. Sørensen used free-hand sketches made with a soft pencil in most cases, and only in a few instances did he use drawing instruments. The thickness of the line remains constant no matter what it represents. The forms are drawn only in outline and are shown in contour only. Sørensen used no shading or other technique to indicate relief or movement in the terrain. The site plans are thus are almost naïve in their simplicity and appear quickly conceived and drawn. To make them more legible and effective, the designer has introduced short, typewritten descriptions at strategic places that reinforce the drawings with words. In the end, the drawings are more analytical than experiential but neatly match Sørensen's intentions in writing the book.

The contrast between drawings by Erik Glemme and those by his contemporary Sørensen could not be more extreme. Sørensen's drawings pared the design down to pure form.[12] It is said that, although he himself was a skilled gardener, he considered only about a dozen trees, shrubs, and herbaceous plants were sufficient for a basic plant vocabulary. To Sørensen, the overriding factors for plant selection were shape and form, texture, and color. Many of his own site plans bear witness to this belief, experimenting with elementary geometric shapes in different combinations. In his youth, Sørensen had the ambition to become either a printer or a gardener, and in his site plans he combined aspects of both trades. This tendency is most evident in the project for the allotment gardens in Naerum, north of Copenhagen, of 1948 [5-6]. The plan resembles a wallpaper pattern, an even pattern of identical forms, seemingly printed with a rubber stamp. Only a few of the ovals in the lower corners have blackened edges; the others are identical prints. Are these biological cells seen through the lens of a microscope? Or a flock of turtles floating with an ocean current? Or even a pattern on a kimono?[13] Sørensen's site plan does not provide much information about the details as such. Then, again, his presentation lies in perfect accord with his design vocabulary: a pattern constituted of simple forms, a restricted palette of plants, an elaborate design applied to an everyday commission. Paradoxically, both Sørensen's and Glemme's site plans would probably work rather well as construction documents. Their presentation types respectively relate in a congenial way to two opposing design attitudes: Glemme's naturalistic plans for projects related to the landscape and Sørensen's abstract plans for exercises in elementary geometry.

THE PERSPECTIVE

A perspective attempts to depict in two dimensions the visual impression of three dimensions. From its Renaissance development and single vanishing point, the constructed perspective has metamorphosed into full-color renderings today most often generated by the computer.

5-5
C.-Th. Sørensen.
Garden study, Denmark, 1966.

C.-Th. Sørensen, Untypical Gardens for a Typical House: 39 Garden Plans

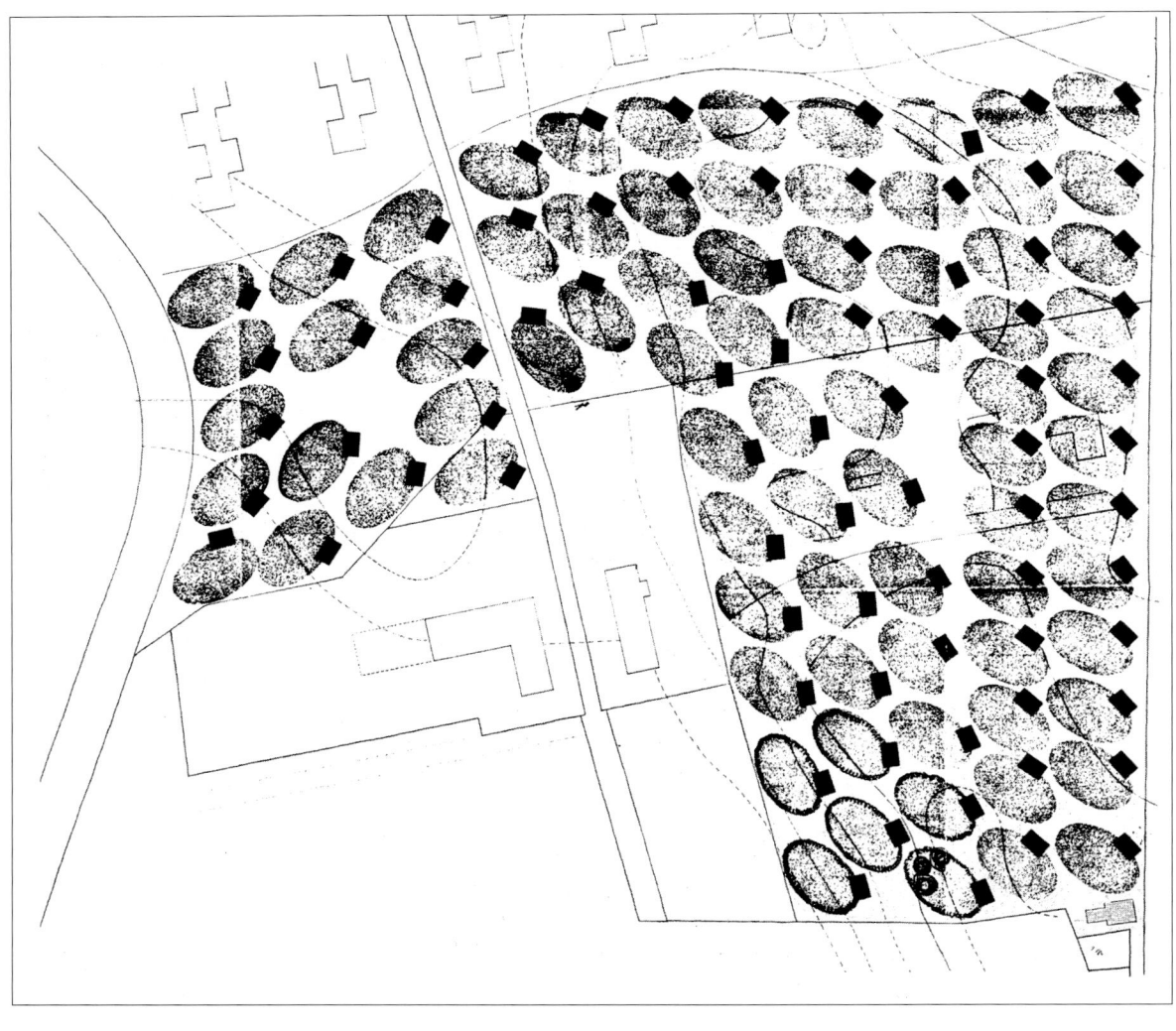

As a free-hand device, perspective can help the designer investigate his or her own ideas, but it then requires a certain skill that not many designers master. As a mathematical construction it is beyond suspicion and never lies, but then again it only depicts the intended reality from one single viewpoint out of millions of possibilities and makes assumptions both about the sense of sight and the shape of the world. These, in turn, make the selection of the station point an argument in itself. Perhaps the perspective has been used above all as a propagandistic instrument with which to persuade, for example, a client. As a computer-generated virtual reality, the perspective achieves a logical perfection but at times becomes the product of programmer's keyboard. These images are often stiff and numb, lacking the poetry of a more imaginative picture, which at its best may induce the feeling of a fresh breeze, swaying tree canopies, the scents of flowers, and the sound of voices.

Walter Bauer (1912–1994) ran an active landscape architecture practice from the 1950s through the 1980s. In his Stockholm office, he employed his wife Lisa to produce the office perspectives. Lisa Bauer, an artist renowned for her own work, drew images for the office's projects with great detail and considerable realism. They are more the work of an artist than an architect, producing the illusion that the project is not only a proposal but already exists.

Although Lisa Bauer often used color in her own art, her landscape sketches were drawn in ink and seldom rendered with additional media. They rank as artworks in the sense that they convey impressions of reflections in water, movement in long grasses, shifting shadows on the ground. The trees are often depicted with traces of a leafy canopy and the pattern of the bare branches simultaneously, thus evoking a sense of time. In the 1975 sketch made for the restoration of Engelsbergs Bruk, the grassy slopes, the building's dark stone foundation and the moving surface of the pond create an associative image with a legible atmosphere [5-7].

Another perspective drawing with similar artistic qualities shows the hermit's hut in the park of Forsmarks Bruk [5-8]. Darkened tree trunks tell us that we are deep in the woods; a huge rock forms the back of the hut with its low door and geometric ornament. We can almost imagine the hermit himself peeking at us from behind a tree. A fruit orchard, intended to complement the garden of Sundbyholm, shows widely planted apple trees in a flowering meadow of *Fritillarias, Aquilegias, Myosotis*, and wild tulips [5-9]. The image should not be taken literally, however; the trees as depicted appear to be at least half a century old and hardly just planted, but the sketch creates an atmosphere of what could come to be in the distant future.

5-6
C.-Th. Sørensen.
Allotment gardens, Naerum,
Denmark, 1948. Site plan.

5-7
Walter Bauer, landscape architect;
Lisa Bauer, delineator.
Engelsbergs Bruk, Sweden, 1975.
The mill pond; perspective sketch.

5-8
Walter Bauer, landscape architect;
Lisa Bauer, delineator. Forsmarks
Bruk, Sweden, 1977, The hermit's
hut; perspective sketch.

5-9
Walter Bauer, landscape architect;
Lisa Bauer, delineator.
Sundbyholm, Sweden, 1977. The
fruit orchard; perspective sketch.

Apart from a few years in his early career, Walter Bauer always ran his own office that came to specialize in restorations of historical gardens. His practice had to compete for commissions and images were used for just that purpose: as a tool to persuade the clients to accept the proposed design. They were a part of the marketing process. With the help of Lisa Bauer's drawings, the office achieved remarkable success and secured numerous commissions over a long period of time. The fact that most of the projects were restorations made these sensitive perspective drawings all the more important. Restoration must demonstrate deep respect for heritage, and often radical changes were not evident in the presentation sketches. Lisa Bauer's perspective drawings emphasize the atmosphere of the design as well as its description, an achievement becoming even rarer today.

Gunnar Martinsson (1924–) spent most of his career as a Professor of Landscape Architecture in Karlsruhe, Germany, a post he assumed at the age of 41. Martinsson ran his own office in parallel to his teaching, designing landscapes in both Germany and his native Sweden. In his early years of practice he frequently participated in competitions, made designs for private gardens, and wrote or co-authored several books—all three activities common for an ambitious young professional on his way to a successful career. Martinsson's active period coincided with the era of modernism and his projects carry many recognizable signs of the times. He often uses elementary geometries, he salutes cutting-edge technique, he looks for novelty, and his design displays a purism that at times borders on austerity.[14] In these aspects he fits well into continental landscape modernism, but hardly into the Swedish modern tradition. In his homeland, modernism in landscape architecture implied quite the opposite, namely, turning to the natural surroundings but leaving the romantic Arts-and-Crafts garden culture behind.

Martinsson's perspective drawings possess a certain quality that embodies many characteristics of modernist thinking [5-10]. They are constructions using their own means, perfect images of a perfect world. There is no sketchiness and everything appears under control. The black lines in ink appear as if etched into the surface of the vellum, every stroke witnessing a clear intention. The drawings are complete, without hesitation or uncertainty. Vegetation appears almost as an exhibition item: identical trees, often without leaves, broadcast the architectonic structure of the branches; hedges are clipped as green rectilinear walls; perennials appear in immaculate studies. All the joints between the paving stones are presented, the furniture is arranged in an orderly fashion, but seldom do people inhabit the images. These perspectives lack transparency: all objects are solid, occluding the features behind them. Every section of the drawings is razor-sharp and no

5-10

Gunnar Martinsson. Garnisonen, Stockholm, Sweden, 1970. Perspective sketch of the office building's central courtyard.

5-11

Gunnar Martinsson, Nordiska Villaparaden (housing exposition), Norrköping, Sweden, 1964. Perspective sketch.

blurriness enhances the sense of distance. There is a certain flatness to these images, in spite of the fact that the perspective as a drawn construction normally heightens the sense of depth. Martinsson seldom uses color or shading, and all the tree canopies display the same level of detail in all their parts [5-11]. The overall impression is that everything ranks with equal importance and that the designer commands it all. Often the pictures have frames around them, forming perfect squares. If a title block is included, it is written by hand with an accuracy that matches typing, all carefully composed and integrated into the finished composition. Nothing has been left out and there is nothing to be added, the designer seems to say.

Sven-Ingvar Andersson (1927–2007) is of the same generation as Martinsson. For a few years they worked together in the office of the legendary Swedish landscape architect Sven Hermelin, who is considered the doyen of Swedish landscape architecture. The two young men consequently shared many things in common that helped them professionally. However, if we compare Andersson's use of the perspective to Martinsson's, striking differences are noticeable, and in some ways they are each other's opposite. Andersson omits things. His drawings want to transmit only the essentials, the most important message he has to deliver. There are few drawn lines in the images, but every one of them has a purpose. They are realized mostly by free hand, drawn directly from the designer's mind rather than from a constructed perspective. The lines have the elegant flow and decisiveness of someone who is already convinced about the efficacy and correctness of the design. In Andersson's case, the perspective drawing is not utilized as a working tool to test the design, but as a final touch that will make the proposal complete. The Rødovre City Hall was designed by Arne Jacobsen in 1956 with a second phase added in 1969. Over time, the austerity of Jacobsen's design for the grounds was experienced as too barren, and in 2000 Andersson was commissioned to improve the situation. His solution was to add a long, rectilinear canal placed parallel to rows of existing lindens [5-12].

The solution respects the spirit of Jacobsen's design despite the strength of the addition. Andersson's perspective sketch indicates rather than shows the purism in the composition of Jacobsen's buildings and landscape. The new canal appears as a calm sheet of water; edges and other detailing are omitted. A bicycle rack in the foreground stands out in almost shocking contrast to the minimalism of the design. The sky mirrored in the water, and the shadows of the lindens, are subtle yet strong effects that the perspective sketch successfully mediates.

5-12

Sven-Ingvar Andersson. City Hall, Rødovre, Denmark, 2000. Addition of a canal to the existing landscape design, perspective sketch.

The Danish architect Arne Jacobsen designed the original project in 1956.

The perspective from Uraniborg is taken from the air. The project involved the restoration of the sixteenth-century Renaissance castle with garden that the famous Danish astronomer Tycho Brahe had built for himself on the small island of Ven. The complex was totally demolished shortly after Brahe's departure in 1597. Old engravings show a castle and a garden created in perfect symmetry, planned according to Leon Battista Alberti's theories about sacred numbers and pure geometry. Andersson's solution was to inscribe the castle's footprint in the ground and to reconstruct only one-quarter of the garden The other three parts are left to the visitor's imagination using the principles of symmetry that structured the original design. It is a project built on one strong and original idea. The perspective sketch shows the scheme but— typically for Andersson—lies somewhere between imagination and built reality as it actually shows more than the intended quarter of the garden restoration [5-13].[15]

Sven-Ingvar Andersson has his origins, education, and early professional experience in Sweden. In 1963, he was appointed Professor at the Royal Academy of Fine Arts in Copenhagen, as successor to C.-Th. Sørensen. Like Martinsson, he has run a practice in parallel with his teaching. Andersson's projects are located mainly in Denmark and Sweden but also in continental Europe. He has written extensively and for a long time has been one of the most acute critics in the field. His drawings share this focused attention and acute presentations of his designs and as such appear equally as critical documents and representations.

CONCLUSION

The relations between the landscape architect, the program, and the client often determine the nature of the graphic representation. Controlling all segments of the process, Holger Blom could afford to work in a pragmatic way.

His idea of building parks using primarily quite loose drawings assumed a careful regard for the nature of the site and an in-house construction crew. Two centuries earlier Fredrik Magnus Piper worked in an era when construction drawings were relatively rare, with detailed design and execution supervised on site. The client, the king, was unrestrained by budget and he often selected his design from artistic impressions conjured by Piper's ink and watercolor renderings rather than from technical drawings. Both types demonstrate how the terrain itself can serve as the basis for a design. Erik Glemme's skilful pencil work produced site plans sufficiently accurate to be used as construction drawings as well as pictorial drawings. The drawings have a sense of authenticity despite the fact that the site plan always shows the projected reality from a theoretical position far above the ground. C.-Th. Sørensen's loose sketches function in a rather different manner, focusing on the idea rather than the detailed design.

The drawings of Sven-Ingvar Andersson and Gunnar Martinsson occupy two opposing poles, the former using perspective to exclude aspects of the project in order to clarify an intention, the latter including every given detail to demonstrate the designer's control. Martinsson depicts the features of the design and their spatial interrelation; Andersson gives us an impression only of his main ideas. Walter and Lisa Bauer use perspective to win the commission and to convince the client of its benefits. All the examples discussed reveal how different modes of presentation reflect the nature of the program and site, the need for communication between designer and client, designer and designer, and designer and builder.

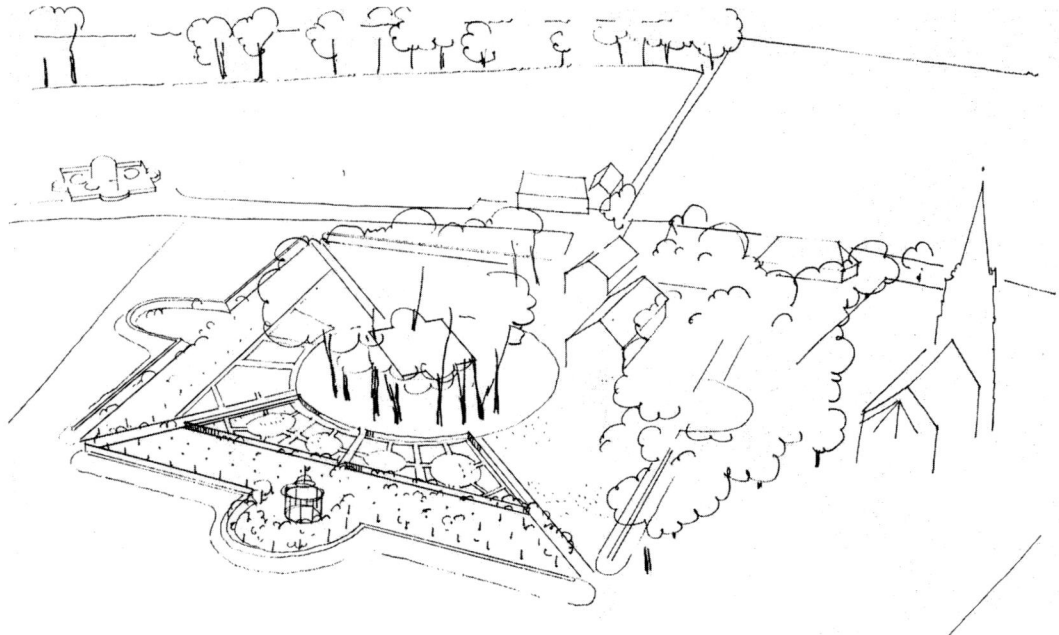

NOTES

1 Chandra Mukerji, *Territorial Ambitions and the Gardens of Versailles*, Cambridge: Cambridge University Press, 1997, p. 170.

2 Munich's Englischer Garten is considered one of the first public gardens in Europe.

3 For a more comprehensive analysis of the period, see Thorbjörn Andersson, "To Erase the Garden," in Marc Treib (ed.), *The Architecture of Landscape 1940–1960*, Philadelphia, PA: University of Pennsylvania Press, 2002.

4 Thorbjörn Andersson, "The Functionalism of Garden Art," in Claes Caldenby (ed.), *Sweden*, Stockholm: Arkitekturmuseet, 1998, p. 238.

5 William Chambers, born in Gothenburg, had spent many years in Sweden during his time with the Swedish East Indian Trading Company.

6 John Harris, *F.M. Piper: His English Garden Studies*, Stockholm: Byggförlaget, 2004, p. 118.

7 Sven-Ingvar Andersson, Margrethe Floryan, and Annemarie Lund, *Great European Gardens: An Atlas of Historic Plans*, Arkitektens Forlag, Copenhagen, 2005, p. 124.

8 Gustav III once called these, "these beloved works of my free hours and the delightful refuge of my worries."

9 Piper said: "the chance of capriciousness, which often leads to absurd odds and ends even in the design of Royal Gardens, if one listens to the conceited Ideas and whims of all the Ladies and Cavaliers." Manuscript, in Fredrik Magnus Piper, *Description of the Idea and General Plan for an English Park*, facsimile, Stockholm: Royal Academy of Fine Arts (1811–1812), p. 154.

10 The majority of the smaller buildings were entrusted to the French theater architect Jean Desprez.

11 Piper's manuscript is called *Description of the Idea and General Plan for an English Park*, dated 1811–1812. It has recently (2004) been published as a facsimile by courtesy of the Royal Academy for Fine Arts in Stockholm, together with a new commentary.

12 *39 haveplaner* is intended as a pattern book. In most of the office projects, Sørensen displays a far more elaborate representation technique, often with bird's-eye views. C. Th. Sørensen, *Utypiske haver til et typehus: 39 haveplaner*, Copenhagen: Christian Ejlers' Forlag, 1984.

13 Sven-Ingvar Andersson and Steen Höyer, *C. Th Sørensen: en havekunstner*, Copenhagen: Arkitektens Forlag, 1993, p. 136.

14 Bengt Isling and Torbjörn Sunesson, "Gunnar Martinsson," in T. Andersson, T. Jonstoij and K. Lundquist (eds), *Svensk Trädgårdskonst under 400 år*, Stockholm: Byggförlaget, 2000, p. 269.

15 Thorbjörn Andersson, "En himmelsk trädgård," *Utblick Landskap*, 2, 1994: 35.

5-13
Sven-Ingvar Andersson.
Uraniborg, Ven, Sweden, 1994.

The restored Renaissance garden dating from the sixteenth century, perspective sketch.

6

Randolph Thompson Hester, Jr.

No Representation Without Representation

In a democracy, the design of the landscape depends on the representation of the public. This public representation forces inventive drawing. Drawing against or for others is substantially different from drawing with or by others. Among their many tools, community designers consistently employ two focused drawing processes: "drawing on your feet" and "designing upside down." They suggest that representative representation relies upon special drawing abilities not common among other design professionals.

By "representative representation" I refer to the way drawing engages the public through grassroots democracy for designing open space, neighborhoods, cities, and regions. This requires representing both the public and the landscape. Face-to-face collaborative drawing provides the political representation. Graphic depictions provide what we professionally call "representing the landscape." The complex combination gives us a special way of drawing: representative representation.

The word "drawing" here includes the depiction of the landscape through a broad range of media, from sketching and painting to modeling by hand or machine. It addresses a great variety of purposes, such as understanding a place, communicating the dimensions or essence of a space, exchanging spatial, philosophical, or programmatic ideas, and imagining choices for changing a site. The entire participatory process, from beginning steps such as active listening to final steps including post-construction evaluation, uses drawing as a central method of communication. Obviously, only a few of these represent the landscape in a literal, figurative way, but all are essential to the public design process.

There are five domains especially critical to democratic landscape design:
1. Representing people.
2. Coauthoring design.
3. Provoking the familiar and the strange.
4. Nurturing stewardship.
5. Empowering people to represent themselves.

REPRESENTING PEOPLE

The Civil Rights Movement and related urban renewal and freeway battles made designers painfully aware that we did not possess the skills to adequately represent people in the design process.[1] In the worst cases, people were ignored altogether, depicted solely as objects in Cartesian space, or as a standardized normative "everyman." Certainly Negroes, the poor, the elderly, or the slightly deviant were not represented at all.[2] The recognition

6-1
Dana Park, Cambridge, MA, 1968–1975.

Moving seating to the edges of the park reduced conflicts, allowing turf control and created a large multiple purpose open space that gang members share with all other users.

Community Development by Design, Berkeley, California [hereafter CDbD]

of this problem forty years ago led to a concerted effort to understand and portray human perception and cognition of, and response to, both the urban and wild landscapes.[3] Over time, sociologists and environmental psychologists have produced a substantial body of research on these subjects.[4] In retrospect, it seems that their findings, expressed in visual and spatial terms, were most used by designers; less imageable research, no matter how important, remained unused. That is the likely explanation why Kevin Lynch's work on perception remains more familiar today than subsequent research on the topic.[5] Drawing techniques, like behavior mapping or social ecology, require careful observation to create patterned visualizations, not unlike soils maps or vegetative mosaics with which landscape architects are accustomed.[6] In a classic discovery of the utility of territoriality mapping, designers plotted the turf of the Dana Park gang in Cambridge, Massachusetts, which explained in spatial terms conflicts with, and crime against, other users of the park [6-1, 6-2A, 6-2B]. These ultimately instigated the primary analysis that inspired a park design that solved what had been chronic turf wars.[7]

As designers discovered when producing a town plan for Haleiwa, Hawaii, the same careful sketching by a participant observer can uncover patterns of sociopetality, idiosyncratic behavior, and social interactions prompted

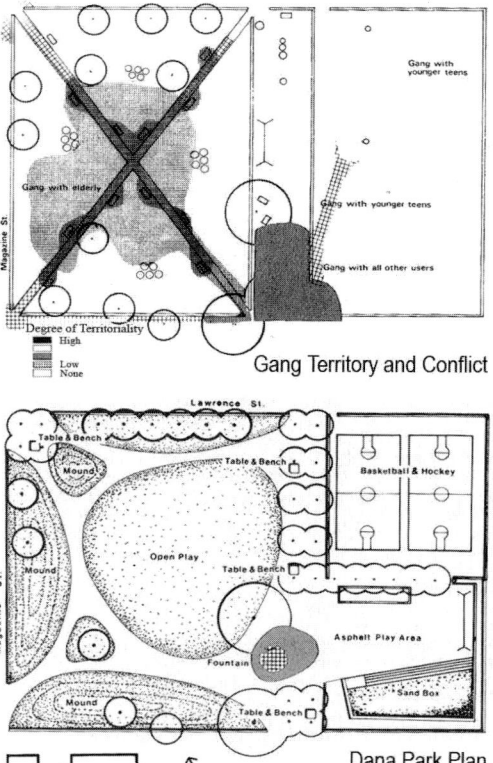

Gang Territory and Conflict

Dana Park Plan

by environmental stimuli. Painting is the preferred medium because it forces designers to observe more carefully during drying breaks. As the designers sketched important spots and events—for example, Matsumoto's Shave Ice, luaus, and the Ice House—recurring patterns of social centering emerged at the interface between indoor and outdoor spaces, patterns tempered by changing sun patterns. This sociofugality was unconsciously designed in the vernacular landscape; planners employed the same pattern to create more such places particular to Haleiwa's climate and culture.[8]

Both the Dana Park and Haleiwa projects rely on the observation of behavior in space. In the former, technical and systematic research methods uncovered noticeable patterns; in the latter, the insight came from the careful looking required by the medium of watercolor. In these and many other cases, spatial patterns from social research and/or first-hand observer drawings allowed designers to better represent the people and to thereafter meet their special needs through landscape design.[9]

COAUTHORING DESIGN

Through experience with participatory design we learned that transactive processes enrich the making of built landscapes, but designers and involved citizens require a mutual empathy and common language.[10] Designers must learn to walk in the shoes of users and vice versa; we had to communicate clearly without jargon. Users had to learn the language of landscape in order to coauthor designs. Drawing has been our most useful language; when thoughtfully executed, a drawing is less ambiguous than spoken language, especially given differences in culture, class, and gender language.

Once, while sitting in a community meeting, I realized I was drawing upside down so that community members could more readily read the ideas we were generating. The utility was obvious, and practice increased my skills. Drawing in this relation to the public is both useful and of symbolic import. Whenever I write upside down I notice that it positively affects the collaborative dynamic because the group is alerted about the seriousness of communicating using a precise and shared language. We must ensure that we understand each other, even if the process requires more time than verbal discussion alone.

And we must be attentive to the various languages that the participants use. For example, several distinct languages were essential to the design of a new community center in Yountville, California. One mother always studied the drawings carefully but said little during the meetings; she would take

6-2A, 6-2B
Dana Park, Cambridge, MA, 1968–1975.

The map of the gang's territory represented the gang, explained the conflicts with other users, and led to a design that resolved the spatial hostility.

CDbD

the drawings home and write letters to the designers describing her imagined use of the spaces proposed; she then suggested detailed ways to improve the design. Her written language was spatial prose, insightful, and precise. The city manager's language was the capital improvements' budget spreadsheet, architecturally graphic, but hardly spatial. The designer's measured drawings represented another visual language and they evolved from the sketches and plans. Each participant had to literally and metaphorically write upside down in order to communicate effectively while coauthoring the design.[11]

Sketching the words of another person requires aggressive listening. The designer sketches while listening, trying to give form to the idea the citizen expresses verbally. The resulting sketch tests whether two or more people are visualizing the same idea and as such may become the medium of exchange, a way to elaborate or create new designs. In some cases, the sketch may replace verbal communication. The construction of Marvin Braude Park in Los Angeles involved rebuilding parts of the Santa Monica Mountains scarred by a failed freeway. Where mountaintops had been removed, thousands of cubic yards of soil from nearby mudslides were used to recreate the original topography. Although layout and grading plans provided a general direction, most of the detailed design, including earth form and rock placement, was undertaken in the field. The contractor responsible for the earth work and the designer learned that words, and even flag markers, would not produce the desired results. In response, the designer sketched as he spoke, starting by directing the grading to landmarks via sketches, and at times using them to guide the bulldozer that followed. Sometimes the contractor took the designer's sketchbook and redrew the most likely outcome of a slope stabilization or drainage way. The earth mover and landscape architect soon began drawing perspectives of the next day's—and even week's—work with locations triangulated to reference points in the landscape, in some cases, miles away. Eventually these field drawings determined almost all the decisions about the pathways and overlooks; quick perspective sketches prevailed, over more formal working drawings, which became of less immediate value.[12]

Sketching is a convenient medium for communicating between two people. Although equally useful, it is far more difficult to use sketches with large groups of citizens. I have observed facilitators, such as Daniel Iacofano, drawing their ideas on butcher paper almost as fast as a hundred citizens could generate them. Few designers can do this, yet it is a critical skill for democratic design. With training and practice, other designers have learned to listen, draw, exchange, and even paint with sufficient precision not just

6-3
Parque Natural, Los Angeles, California, 1998–2003.
The plan for the Natural Park was co-authored under a tent on site.
CDbD

to record ideas, but to design with large groups. When planning Parque Natural in South Central, Los Angeles, landscape architects designed most of the project on large tables set under a tent on the then-derelict site [6-3, 6-4]. They drew the organizing principles while community members and other design team members discussed critical issues. The form of the main design features, including a community center and *zócalo* (plaza), recreated arroyo and wetland, natural paseo (walkway) and meadow, were coauthored using interactive sketching. Details of the architecture, including the dimensions of the columns needed to create social spaces at the entries, lighting for the meeting room, and the materials of the floor pavers, were negotiated by quick painting. Idea. Sketch. "Do you mean this?" "No, more diffused light." Sketch. "Like this?" "Yes, that's better." This transparency of the design process elicited creative exchanges.

The contentious issue of fencing was innovatively resolved through the sketch process. Due to gang warfare and general safety concerns, residents listed park rangers and fencing at the top of their program requirements for the park. Staff and some community leaders were opposed to fencing this site largely because they assumed it would be built of an unwelcoming chain link. The residents insisted, however. During one debate, a designer remembered an exquisite ironwork fence he had seen in Spain. As people argued, he sketched it from memory [6-5A, 6-5B]. "What about doing a fence something like this?" In the group were employees of the numerous metal fabrication industries in the neighborhood, one of whom responded, "We can make that." Another moved closer and said in Spanish that it couldn't be made as drawn. The designer couldn't understand him, but that hardly mattered because he was already correcting the drawing by proposing a detail that was easier to manufacture. Over the course of succeeding design workshops, they resolved issues of liability, building standards, cost, and details. The design evolved from fuzzy, abstract memory to a fence picturing a metal wetland with marsh reeds and native egrets on its gates. With misunderstood words as its background, the coauthored sketches shaped the park.[13]

PROVOKING THE FAMILIAR AND THE STRANGE

Teaching citizens elementary professional spatial drawing skills can produce significant public design benefits. All citizens can map and draw, some quite well. But, like beginning landscape design students, they need to be assisted in observing the landscape carefully, thinking complexly, imagining nontraditional resources, using specific precedents, accounting for natural changes in the landscape, generating spatial concepts, and evaluating plans.

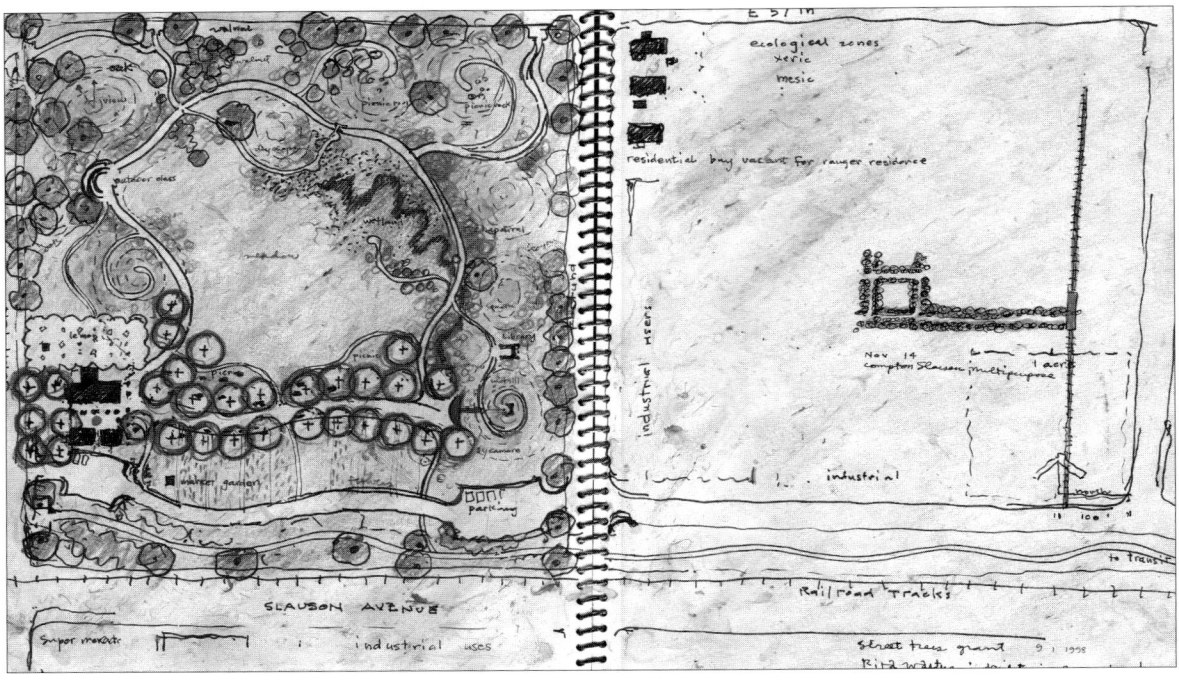

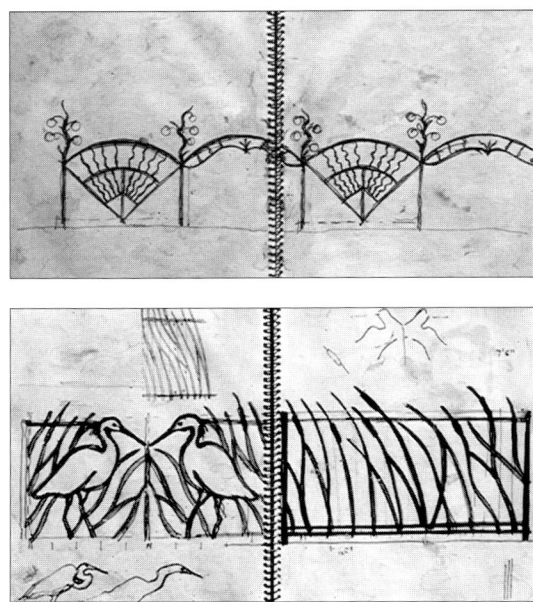

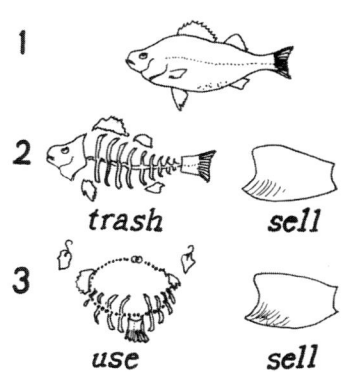

6-4 [opposite above]
Parque Natural, Los Angeles, California, 1998–2003.

Everything but the working drawings was sketched on site, mostly upside down so that community members and designers could share in the design process.

CDbD

6-5A, 6-5B [opposite below]
Parque Natural, Los Angeles, California, 1998–2003.

Fencing was created by a back-and-forth process; ideas were connected by local craftsmen until handsome and easy-to-fabricate details emerged.

CDbD

6-6 [above]
Fish heads.

To provoke creative thinking, the familiar is made strange and the strange familiar by looking for undiscovered resources such as fish heads in the community. A diagram clarifies the intent and expected outcome.

CDbD

In public landscape design there is often a rush to judgment without a careful consideration of the place. At times, community leaders make decisions without ever visiting the site, although we know that meditative observation often reveals nuances key to a design that captures the unique character of that place. Participatory exercises that require discriminating looking can enrich the public discussion. Most residents stated that they had never gone to Runyon Canyon in Hollywood because they feared for their safety, and a momentum was building to remove all the vegetation. The designer asked residents to note the exact spots they feared and why. The resulting Fearful Places map showed several locations with concerns stemming from the presence of homeless people and long, narrow walkways. Residents felt completely safe in over 90 percent of the site, however. The drawing exercise focused attention on the trouble spots, prevented indiscriminate vegetation removal, and led to revegetating with native chaparral and saving an old botanical garden.[14] Urban landscapes must be experienced sensually to be appreciated and understood. On-site sketching workshops encourage this engagement.

In all poor communities—and most comfortable ones—open space development depends on discovering some previously unrecognized resource upon which to capitalize the project. This process is called "finding fish heads," that is, a means by which a waste product can be made useful [6-6]. Communicating the idea of fish heads to community members prompts them to seek and find uses for the seemingly useless. They may photograph, sketch, or simply list potential fish heads. But often the designer must map and sketch the possibilities before they are recognized.

In the design of San Vicente Mountain Park in the Santa Monica Mountains of California, the design team recognized that the remnants of an old Nike Missile base, if restored, could encourage historical interpretation and wildlife viewing. To the citizens, the existing tower was a safety hazard, the concertina wire fences were inappropriate in a park, the concrete bunkers were eyesores. During a design workshop on-site, however, quick sketches transformed the tower into a wildlife observation deck, the walkways enclosed with concertina wire into interpretive trails, and bunkers into benches [6-7]. The recycled military artifacts provided a special place, capturing the past while reusing the missile-era relics to satisfy desires for nature study, picnics, and mountain biking.[15]

Although most people don't possess a knowledge of great historical landscapes from four years of history courses, most have a storehouse of personal

experiences that provide qualitative and quantitative references useful in public design. However, people generally have a hard time recalling these precedents with sufficient detail for use in the process of design. Self-hypnosis, or guided fantasy, can help recall and draw a favorite street, square, or other landscape with sufficient detail to inform discussion. Drawings made after a hypnotic visualization are amazingly precise spatially while also capturing more ephemeral aspects such as light quality. These drawings allow citizens to be much more critical and realistic about the size of a site, what would and would not fit, when a site becomes too crowded, or the impact of a low versus a high tree canopy. The informed discussions that result could never develop without the detail of precedents visualized and drawn. This method is particularly useful in determining what people mean by "natural," a key concept for landscape design and one with many abstract interpretations. Drawings make the abstract concrete and guide design.[16]

Visualizing natural change over time, whether old-field succession or park vandalism, is one of the most challenging aspects of landscape design. Drawing the expanding shadow from a maturing tree, or producing an impression of a place from a century ago—or what it will be like a century from now—helps the designer or lay person create landscapes distinguished by regional forces.

6-7

San Vicente Mountain Park, Santa Monica Mountains, California, 1995–2000.

The derelict military tower was a safety hazard to some, a fish head resource to others. It was transformed into favorite city overlook and wildlife observation deck after drawings showed its potential.

CDbD

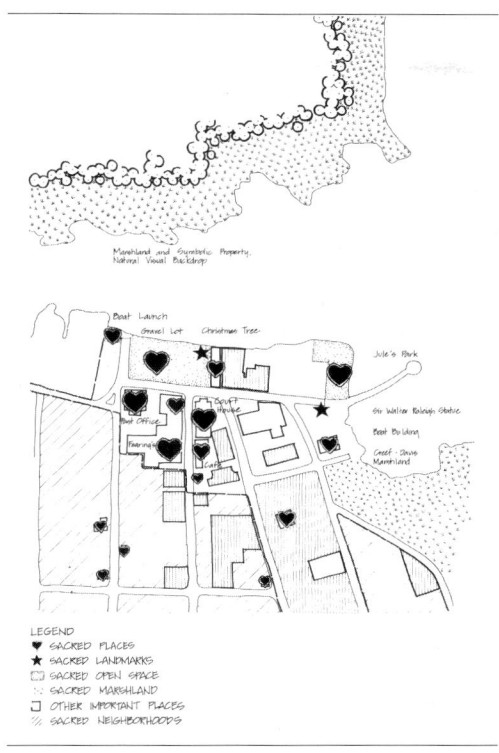

In Castle Rock, Washington, a city partially destroyed by Mount Saint Helens,
a design team worked with long-time residents to produce large pastel
drawings of the Cowlitz River passing through the town before the eruption,
immediately after, ten years after—and how it might appear in the future.
At the time of the workshops the site was a dead wasteland of gray pumice
and debris, in places 50 feet high. Some years before, however, it had been
a maturing riverine ecology teeming with wildlife. When the drawings were
shown at a community meeting, one woman looked at the lifeless present
and said: "The birds don't sing here any more." The audience hushed, many
cried. It was no overstated metaphor. The songbirds disappeared when
their habitat was washed away or buried by the debris flow. The drawings
of the future state offered hope—but only after the catharsis instigated by a
set of evolutionary drawings distributed throughout the city.[17] Understanding
the changing nature of the landscape via drawings is always useful, but seldom with such dramatic results.

Citizens become more effective partners in designing landscapes when they
are deeply rather than superficially engaged in the problem solving. One
of the most critical aspects of landscape architecture is achieving a gestalt,
the essence of a place that cannot be derived by adding up the various

characteristics. The Mexican architect Ricardo Legoretta once said that getting the essential idea is 95 percent of design. It is not additive, or even qualitative, and certainly not easy to draw—even for the most accomplished designers. But citizens have helped designers visualize some of the most profound landscape concepts using a combination of sketching, collage, poetry, and the free association that follows rigorous analysis. In Manteo, North Carolina, after analyzing a number of collages expressing what citizens considered their gestalt, the mayor expressed the concept of "Come sit on our front porch, let us tell you of the dreams we keep." The front porch provided both metaphorical and literal inspiration for the community plan, one that has received multiple design awards and been extensively published [6-8].[18]

Evaluating plans prior to construction is an important aspect of public landscape architecture. Using methods described above, most notably, drawing activity and social ecology patterns, citizens can complete these evaluations. Citizens as potential users, however, must imagine (rather than observe) how they and others will inhabit a space. The resulting drawings may correct design flaws before construction, saving money and enhancing social suitability.

From this cursory review it is clear that drawing engages participants throughout the design process and can be used democratically to address issues critical to landscape architecture. In all the above examples, drawing exercises encourage citizens to more capably contribute to landscape design by thinking from two viewpoints: that of the user and that of the designer.

NURTURING STEWARDSHIP

Increasingly, designers recognize that lasting landscapes depend upon stewards, people who nurture and maintain places over time. Historically, the image—sketch, painting, or photograph—has cultivated heightened awareness, scientific understanding, and the active engagement in landscape preservation. Three techniques employ those approaches in ways in which citizens participate in making the images that lead to landscape stewardship: (1) recording sacred landscapes; (2) mapping citizen science; and (3) making weird science spatial.

In a society seen as increasingly disassociated from natural phenomena and places, recording "sacred" places provides an antidote to environmental anomie. Sacred here describes places that are most meaningful to community life, both everyday and ritual events. The exercise begins by asking people to individually list the places they most value in their city; then in small groups they visit, map, and record those places through photographs or sketches. The exercise typically heightens awareness of subconscious attachments to, and dependence upon, the landscape. The results can be used directly in the community design process, and these methods were first employed in Aurora and Manteo, North Carolina, and in other communities thereafter.[19] The Environmental Protection Agency realized that sacred landscapes mapping could encourage better land management, and in 1997 the agency undertook demonstration projects in Pennsylvania, Maryland, and Virginia to assist communities in identifying and visualizing their sacred places. The goal of one project was to stimulate local stewardship in an effort to improve water quality in the Chesapeake Bay, miles away.

Today, stewardship programs of all sorts—from air and water monitoring to valued places and wildlife corridors—employ mapping by lay people as a central tool. Often these programs, called citizen science, teach basic scientific research methods and mapping. Wildlife stewardship is especially popular among volunteers. The Nature Mapping Program in Washington State, for example, trains volunteers to identify wildlife species and to map the locations of their sightings. Some 50,000 volunteers participate in this program to create a database of wildlife concentrations that supplements professional studies. These are used for making decisions about land use planning.[20]

One of the unique contributions made by landscape architects to conservation biology and habitat design translates complex research into spatial patterns. At the simplest level this translation involves mapping the territories of species to determine, for example, the impact of habitat loss and the core areas and corridors needed to preserve the cougar population in the Santa Monica Mountains in Southern California. At a slightly more complex level, the study established through drawing the spatial relationships among multiple species like the cougar, the coyote, and the quail. Human activities further complicate these relationships, producing island and edge effects; interior species become locally extinct and boundary species increase. In some cases this trend triggers rescue effects, whereby declining species are replaced near urbanization if there are large core habitats in the vicinity. Scientists seldom synthesize multiple research findings into spatial terms, thus the designer plays a role of interpreting and mapping nonspatial research results. In creating the Big Wild in Los Angeles and the Green Plan in Tainan County, Taiwan, designers worked with citizens and scientists to graphically record research (much of which seems counter-intuitive due to the intricate food chain and habitat relationships) in forms useful for planning and design [6-9]. For example, a drawing of the island effects on wildlife of the proposed

Reseda-to-the-Sea Highway was central to creating the Big Wild in Los Angeles. This eventually led to the abandonment of plans for a new freeway and the initial land acquisitions for a greenbelt surrounding Los Angeles. In Taiwan, similar drawings led to the creation of a four-county National Scenic Area called for in the Green Plan. This plan is based on explicit spatial requirements for one of the most endangered birds in the world. Habitat needs, detailed in studies of water depths and foraging range, were drawn by local fishermen, scientists, and the designers to provide the basis for the plan.[21] Visualizations of wildlife, natural processes like runoff and flooding, hungry water erosion caused by releases below dams, and erosion, or nutrient cycles, remain mysteries until clearly diagrammed graphically. When these relationships are simply and accurately diagrammed, citizens possess the basis for ecologically sound stewardship activities like habitat restoration, species reintroduction, storm water management, and urban vegetation enhancement.

EMPOWERING PEOPLE TO REPRESENT THEMSELVES

If used collaboratively, most of the previously discussed drawing techniques develop citizen skills and control. Several techniques are particularly empowering: building community, choosing, and drawing everything essential. Participatory design builds community. Group graphic techniques developed by the landscape architect Daniel Iacofano consciously and effectively achieve this. By visualizing and recording each person's ideas about a design problem on a large sheet of paper, by consciously organizing the ideas by topic, by showing relationships between ideas, and by highlighting areas of agreement, Iacofano creates a shared experience that serves as a collective memory of a design process and heightens a sense of community. As a result, many previously disjointed communities come together to achieve a common goal.[22] Although not as dramatic as the group graphic exercises described above, most collaborative drawing strengthens the sense of community among the participants if everyone is acknowledged and represented fairly in the process.

Drawings that give citizens choices empower them. Visual preference tests and simulations provide citizens with clear visual alternatives in matters of urban and landscape design.[23] Providing choice requires drawings that are easily read, evaluated, and compared. Drawing style must be consistent. The designers' favorites must be drawn no more compellingly than the other choices. The distribution of the choices must be broad, thereby providing a forum for public discussion.

The citizens involved in participatory design learn the power of the drawing in its many dimensions. And they may learn that you can draw almost any-

6-9

Black-faced spoonbill geometries, National Science Area, Republic of China, 1997– 2004.

Compiled by local fishermen, scientists and designers, these detailed diagrams provided the basis for a plan that saved over 20,000 fishing-related jobs and the rare spoonbill from extinction.
CDbD

6-10

When provided with appropriate drawing skills, lay people can build their own communities, make critical choices and ensure that essential factors are visualized precisely.
CDbD

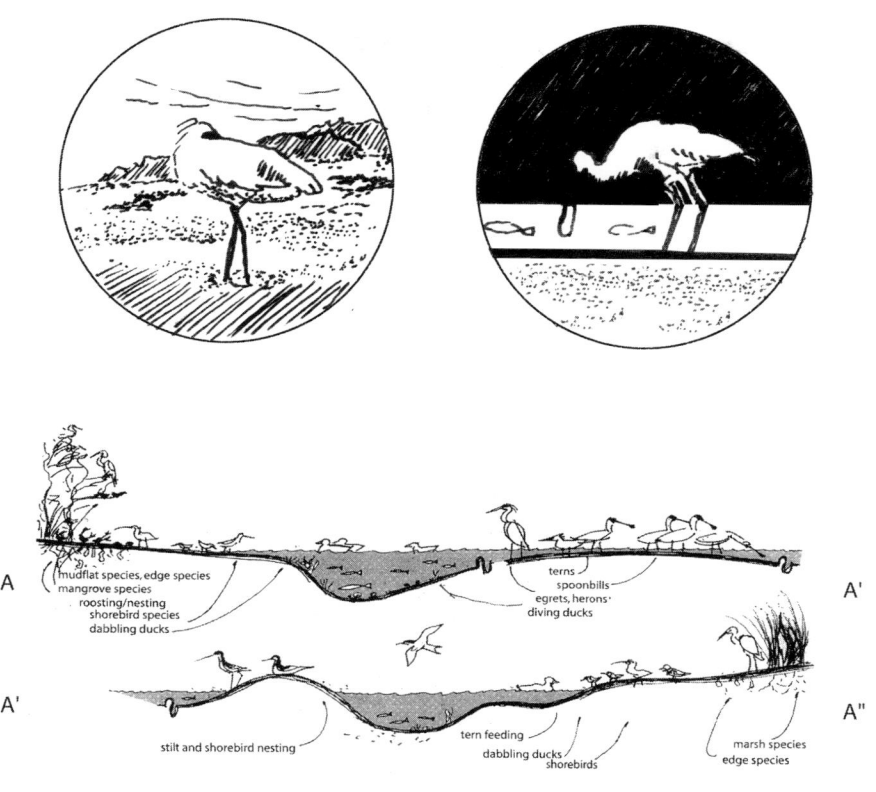

thing—an idea, a place, or an action—and that you can understand it better through drawing. Drawing an unseen thing, whether a map of environmental injustices or barriers to civicness, make that thing concrete and manageable. Abstractions, like power structure or concepts of naturalness, become more understandable when recorded graphically. Drawing things as mundane as preferred picnic settings makes that idea legible and known. In this later case, participants are always shocked at the degree of difference envisioned for the same simple activity. The challenge for the designer is to get the citizens to draw everything essential to developing a plan.

CONCLUSION

Fruitful democratic design depends upon representative drawing, and representative drawing often requires coauthors, often the citizens themselves. Certain techniques discussed here are not unique to democratic design or landscape design [6-10]. For example, securing a gestalt, careful observation, and thinking complexly are essential to design in all disciplines and forms. Other techniques are more particular to landscape design, notably evolving landscape, imaging fish heads, recording sacred places, and making science spatial. Most of these drawing techniques, however, have particular relevance to participatory design, although locational mapping, plan drawings, diagrams, quick perspective sketches, and before and after overlays, are common to most design practice. What differs, for the most part, are the specific ways the techniques are used to convey the importance of what others think, and to say "I want to communicate clearly with you"—to create a common language for complex publics, to nurture an informed civic debate, and to include the excluded. Equally important is to understand the subject as the public perceives and values it. The democratic designer must understand both the technical limits and the cultural constructs such as the experience of nature, ecological science, natural processes, and sociopetality. Then he or she must imbue each drawing with an understanding of function, dimension, materials, and economy. The differences in the intentions behind these drawings are profound. The democratic designer must be able to draw on his or her feet, drawing transparently, quickly, imaginatively, and with professional mastery of the subject at hand.

NOTES

1 Simpson F. Lawson, *Workshop on Urban Open Space*, Washington, DC: US Department of Housing and Urban Development, 1968, pp. 35–36.

2 Karl Linn, "White Solutions Won't Work in Black Neighborhoods," *Landscape Architecture*, 1968: 23.

3 Suzanne Keller, *The Urban Neighborhood*, New York: Random House, 1968, pp. 19–86; R.B. Litton, *Forest Landscape Description and Inventories*, Berkeley, CA: USDA PSW-49,1968; Donald Appleyard et al., *The View from the Road*, Cambridge, MA: MIT Press, 1964.

4 Herbert J. Gans, *People and Plans*, New York: Basic Books, 1968; Harold M. Proshansky et al., *Environmental Psychology: Man and His Physical Setting*, New York: Holt, Reinhart and Winston, Inc., 1970.

5 Kevin Lynch, *The Image of the City*, Cambridge, MA: MIT Press, 1960.

6 Albert Rutledge, *A Visual Approach to Park Design*, New York: Garland, 1981; Clare C. Marcus, *House as Mirror of Self*, Berkeley, CA: Conari, 1995; Mark Francis et al., *Community Open Spaces*, Washington, DC: Island, 1984.

7 Randolph Hester, *Neighborhood Space*, Stroudsburg, PA: Dowden, Hutchinson and Ross, 1975, pp. 54–55.

8 Community Development by Design, *Haleiwa Town Plan*, Honolulu: Haleiwa Main Street Program, 1991, pp. 7–27.

9 Randolph Hester, *Planning Neighborhood Space with People*, New York: Van Nostrand Reinhold, 1984.

10 Paul Davidoff, "Advocacy and Pluralism in Planning," *Journal of the American Institute of Planners*, 1965: 331; David Godschalk, "Collaborative Planning: A Theoretical Framework," in *Eleven Views: Collaborative Design in Community Development*, Raleigh, NC: School of Design, 1973, p. 19; John Friedman, *Retracking America: A Theory of Transactive Planning*, Garden City, NY: Doubleday, 1973.

11 Community Development by Design, *Yountville Town Design Plan*, Yountville: Town of Yountville, California, 1997.

12 Randolph T. Hester, Jr et al., "Road Kill Yields Big Wild," *Proceedings of American Society of Landscape Architects*, 1996: 82–85.

13 Kim Sorvig, "The Wilds of South Central," *Landscape Architecture*, 2002: 66–75.

14 Community Development Planning and Design, *Runyon Canyon Master Plan*, Los Angeles: Santa Monica Mountains Conservancy, 1985.

15 Community Development by Design, *Restrictive Use*, Los Angeles: Santa Monica Mountains Conservancy, 1996.

16 Randolph Hester, "A Womb with a View," *Landscape Architecture*, 1979: 475–481.

17 Community Development by Design, *Castle Rock Riverfront Plan*, Castle Rock, Washington: City of Castle Rock, 1989.

18 Tom Zarfoss, "In Search of Excellence," *Landscape Architecture*, 1984: 44–57.

19 Randy Hester, "Subconscious Landscapes of the Heart," *Places*, 1985: 10–22.

20 Victoria Chanse and Randolph Hester, "Characterizing Volunteer Development in Wildlife Habitat Planning," *CELA Groundwork*, Syracuse, 2002, pp. 25–28.

21 Randy Hester et al., "Local Community Design and Platalea Minor," *Conservation of Black Faced Spoonbill in East Asia*, Seoul, pp. 194–210; Randy Hester, "Big Visions and Little Neighborhoods: Participatory Design in Creating the Los Angeles Greenbelt," *Community Architecture*, Taiwan: Kachsiung, 2003, pp. 93–114.

22 Daniel Iacofano, *Meeting of the Minds*, Berkeley, CA: MIG Communications, 2001.

23 Peter Bosselmann, "Times Square," *Places*, 1997: 55–63; Anton C. Nelessen, *Visions for a New American Dream*, Chicago: APA Planners Press, 1994.

7

Marc Treib

On Plans

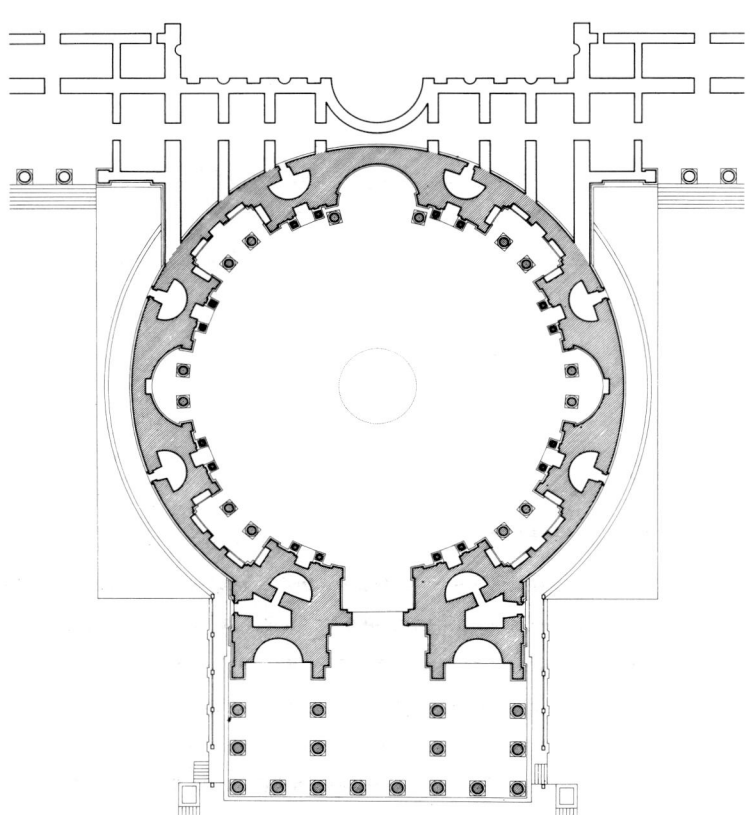

On a certain day during my college years, one of my architecture teachers showed us an amazing document: *Building Footprints*, a portfolio of plans of notable buildings prepared by Eduardo Sacriste.[1] We were told to carefully examine each of the plans—all drawn to the same scale—to understand the logic that underlay the design of these structures, and to comprehend the relationships of their parts. The drawings were magnificent, each of them beautifully rendered in ink [7-1]. The outlines and hatching were consistent; the identifying text was not found on the portfolio plates, but in an accompanying book. My fellow students and I spent hours looking at these drawings, marveling at the elegance of Frank Lloyd Wright's 1938 Johnson's Wax office building, or being amazed at just how small Le Corbusier's 1952 chapel at Ronchamp was when compared to Notre Dame de Paris or even the Pantheon. The lessons taught by that portfolio of plans were indelible, and from that time on, I regarded the plan drawing with a respect approaching reverence.

In English, the word plan—like the word design—can be read as either a verb or as a noun. As a verb, it stands for a purposeful activity, an action with an intended result. We plan a settlement or we plan a garden; we plan a city or we plan a region. As a noun, the word plan suggests a directive document that serves as a guide for some action in the future. More specifically, in the environmental design disciplines, the plan illustrates the relationships among component parts as well as the total result of those relationships and parts in amalgamation. The plan thus embodies the design idea; it is the kernel from which the design develops. It is a scheme, a pattern, the generating force behind the making of a landscape. Or, that is to say, it can be the generating force.

Some landscapes are planned using process as their basis and have no need of any formal directives. Other designs rely on the plan as a scaled representation that guides the making of the landscape. This difference in idea also distinguishes gardening from a garden. To landscape makers such as the nineteenth-century Englishman William Robinson, the garden was the product that resulted from acts of gardening. In books such as *The Wild Garden* (1870), Robinson spoke little of the totality of the garden that would issue from the native or exotic plants of which he wrote; he was more concerned with the process through which the designed landscape was first made and evolved over time. His mission, if not his precise lesson, was restated and enlarged by Ian McHarg in his lectures and publications, including his landmark book *Design with Nature* of 1969.[2] McHarg showed no interest in designing landscapes in accord with any formal plan; in fact, he decried the practice. Instead, he insisted on planning the landscape, in this case, basing his design on an interpretation of ecological processes.

7-1
Agrippa. Pantheon, Rome, Italy, 25–27 CE.

Eduardo Sacriste,
Building Footprints

The incremental application of process can result in grand landscapes that delight the senses and provoke thought, but one is hard pressed to name any of them. But this is not to say that it is impossible to use a process for aesthetic goals. For example, we may intensify the aesthetic aspects of an existing landscape using observation and verbal instruction rather than drawings. In fact, it would have been quite difficult to use plans to make a garden like Stourhead because the landscape developed in space rather than on paper, and over time rather than at one moment. Henry Hoare and his collaborators began with the existing conditions but dammed streams, modeled landform, and even moved a village to bring the landscape into accord with an idealized image drawn from idealistic notions of classical antiquity. From memory, we could easily sketch the position of Stourhead's great house, its lake, Pantheon, and bridge, but only in very general terms could we make a general plan of the garden and its topography. We would probably lack any sense of the dramatic rise and fall of the land, nor the spatial positions of the architectural elements, for a plan tends to be a two-dimensional compression of a three-dimensional experience (some would argue four-dimensional, if we include time). The plan at best is a very abstract(ed) conveyance of a design idea.

Other gardens, such as Vaux le Vicomte, are well served by their presentation plans. In them we sense the presence of geometry in both the drawing and on the ground; the plan embodies the idea of the garden and the garden is a drawing rendered large and volumetric. The relation of château to axis, of planting to parterre and parterre to circulation, of bosquet to terrace—all of these are as clear (perhaps clearer) in the plan as they are when perceived on site. The plan works well to represent the French formal gardens because

their idea requires an understanding of the complete entity, and because the landscape is basically flat. Of course, the visitor to the site would be plagued by problematic features that remain unseen during the initial promenade: Vaux's cross-axial canal comes as a great surprise in space although it is readily apparent in the drawing, but in plan the effect of the distant hillside rising markedly toward the monumental statue of Hercules is flattened and thus diminished [7-2].

A plan presents a view that never exists in reality. It is a convenient fiction, and like the section and elevation, a fiction that denies human binocular vision, saccadic visual reception, the curvature of the earth, and the effects of atmospheric distortion. It is a utopian form of representation that like an X-ray reveals to us relationships normally unseen. For the most part, this is less true for landscape architecture than architecture, where walls and ceilings and roofs conceal the architectonic relationships from view. Only in a high level flight over a ruin would we ever see a place truly in plan. Plans lie beyond normal experience, and this is the very reason so many people have difficulty in understanding them. The plan gives us a coherent idea at the expense of spatial experience, and barters human experience for geometric or geographic congruence. Given that most land-scapes are open to the sky, with sloped, stepped, or folded land forms, a garden is more easily seen in plan than a building—and it is easier to read a plan of a garden into reality. But even here, wooded areas and changes in level can almost completely subvert the design scheme suggested in the plan.

Le Corbusier spoke of the plan as generator, by which he meant that in the plan were held the seeds of all further architectural development: "the plan is the determination of everything."[3] "To make a plan," he writes, "is to determine and fix ideas. It is to have had ideas."[4] The structural system, the pathways, the relation of spaces one to the next—these are all guided by and reflected in the plan. But there are dangers to granting the plan a role too serious, for example, when the designer judges the merits of the scheme on the graphic appeal of the design as a pattern. The ambiguity of some graphic projects can never be achieved in reality—the attraction lies in the optical proper-ties of lines and tones and shapes. The presentation drawings for the Parc de la Villette are far more interesting than the park itself, and almost every one of the achievements claimed by the designer are virtually impossible to discern on site. The plan—and especially the exploded axonometric drawing —promised a formal power absent in the realized design. Here, the generating power of the plan was insufficient; we miss the subsequent adjustments to the abstraction of the plan necessary to create space and visual interest.

7-2
André le Nôtre, landscape architect. Vaux le Vicomte, France, c. 1660.

The view from the hillside reveals the cross-axial canal—a feature clear on the plan, but hidden from view on the promenade from the chateau.

Marc Treib

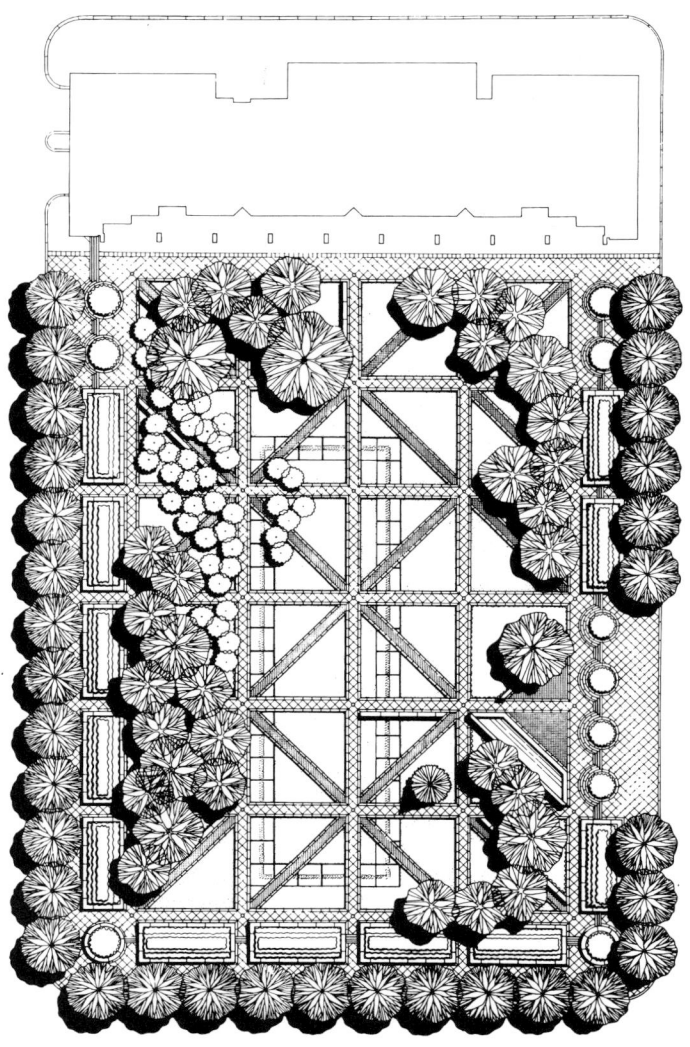

7-3

Peter Walker/The SWA Group.
Burnett Park, Fort Worth, Texas,
1983, Plan.

Peter Walker Partners

7-4

Burnett Park, Fort Worth, Texas.

Marc Treib

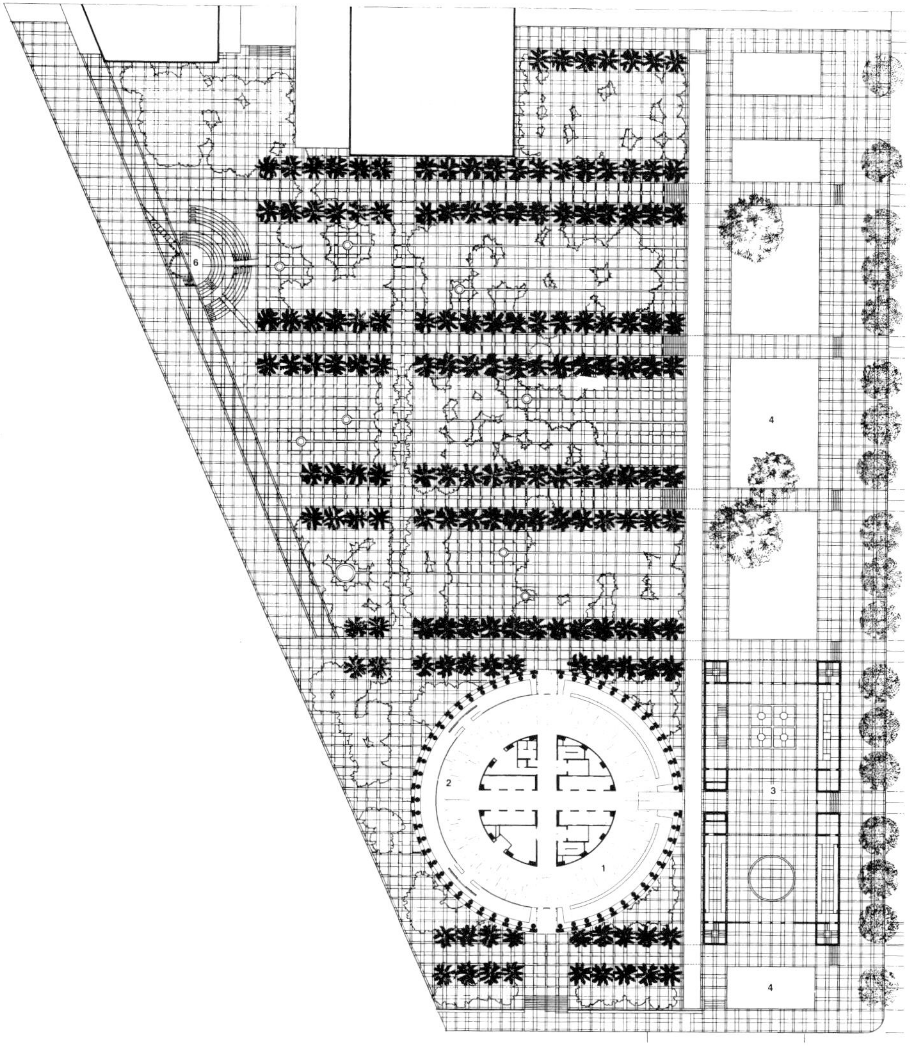

7-5
Dan Kiley, landscape architect;
Harry Wolf, architect, NCNB Bank
Terrace, Tampa, Florida, 1984.
Plan.

Officie of Dan Kiley

7-6
NCNB Bank Terrace, Tampa,
Florida.

Marc Treib

At best, conceptual design uses the plan as a shorthand for complex social and spatial thinking. Unfortunately, in the hands of an inexperienced designer —like many a student—the plan represents the limit of the landscape's or building's design development. To make volumes, the flat shapes of the plan are extruded vertically into three dimensions. In contrast, for an experienced designer like Le Corbusier, the plan represents a condensation of the total design idea grasped cohesively in the mind. As such, it is a type of shorthand that uses the plan as a compressed symbol of something much greater and more complex than the literal reading of the drawing.

In the past decades we have seen many landscapes (and buildings) that sing in their plans but only chant (or grunt) when constructed. The fascination with grids, overlaid patterns, and rotated arrangements of stripes or checker-boards has led to many built landscapes that are interested primarily in their plans. And when the intrigue of that plan can't be perceived (when a building is in the way, for example), or when the project extends far beyond the human scale, the resulting landscape is far less than interesting. Peter Walker and The SWA Group's Burnett Park in Fort Worth, Texas, is stunning as a pattern and as a diagram of projected circulation movement [7-3]. The actual park, however, is less engaging, and little temptation entices the visitor to walk through it [7-4]. Dan Kiley's NCNB Bank terrace/garden in Tampa, Florida, is far more complex and rewarding. Its plan is one of the most stunning of any landscape from this century, a beautiful play of vegetation and paving that Kiley claims to have developed following the numerical sequences of the Fibonacci series [7-5]. To the basic exponential ordering, Kiley added a system of water rills fed by a canal that enriched the ground plane. A grid of Washingtonia palm trees contributed a rhythmic spatial order and architectonic dignity, while swaths of crape myrtle trees set in seemingly non-geometric clumps (though derived from the prevalent geometry of the scheme) countered the regular order of the grid at a lower height [7-6]. The NCNB garden/terrace plan, complex in its two-dimensional patterning, directs spatial composition: it is truly the plan as generator.

The plans of many great gardens surprise us when viewed for the first time; we may even suffer some difficulty in bringing the drawing and the experience into synchronization. Thomas Church's celebrated Donnell garden in Sonoma County, California (1948), is experienced in a far more complex manner than its plan would suggest [7-7]. The use of the splayed enclosing walls (rather than walls set at a simple right angle) and its biomorphic shapes created a spatial experience without direct precedent. The swimming pool—probably the most famous of all modern pools—appears as a free composition whose contour changes with each step. And yet in plan we are surprised to discover that

7-7

Thomas Church. Donnell garden, Sonoma County, California, 1947. Plan.

Environmental Design Archives, University of California, Berkeley

7-8

Donnell garden, Sonoma County, California.

Marc Treib

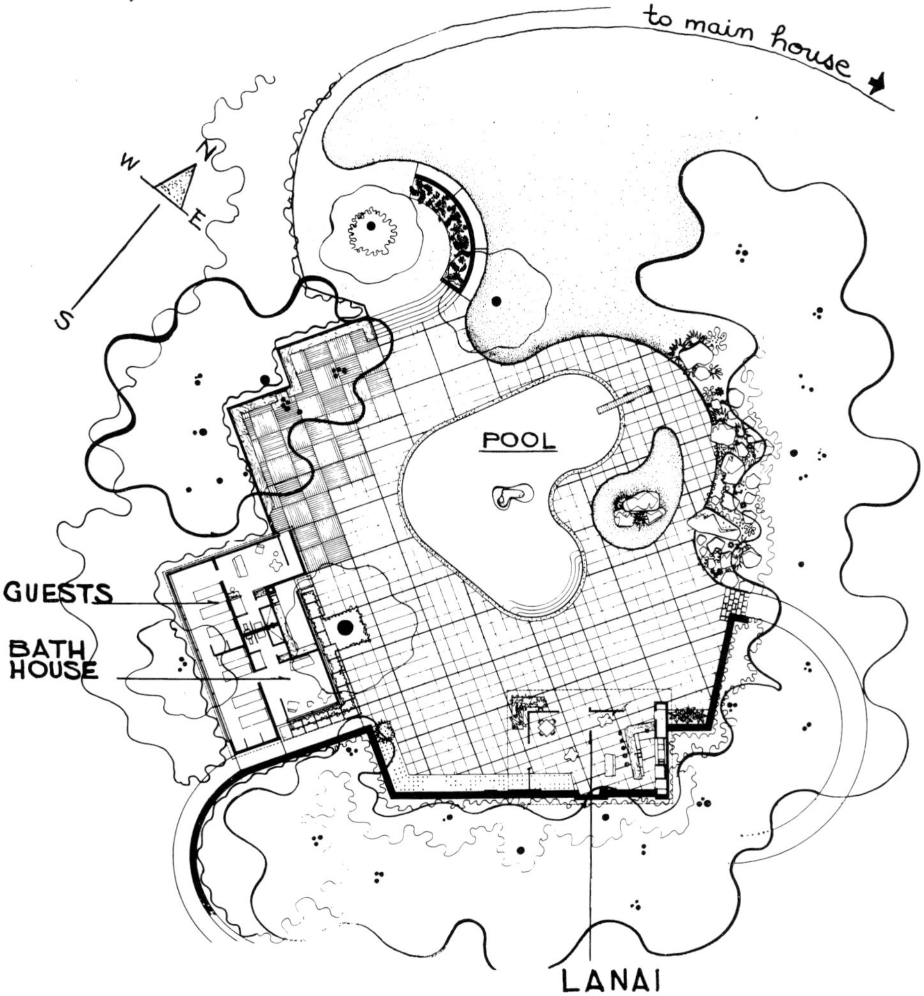

to main house

N
W
E
S

POOL

GUESTS

BATH
HOUSE

LANAI

the geometry of the pool is quite regular, governed by the calculable arcs of the compass rather than the irregular curves of the template [7-8]. In many of Church's gardens, in fact, the reading is richer than the shape, which is often quite simple. To some degree this enrichment results because we ultimately experience each element only in relation to other elements, never in isolation. In addition, the garden's designers only suggested many of these curves in drawn studies; they were fixed only on site—often using a convenient garden house to trace out the contour of the paving or the planting bed.

We have plans of the Katsura Villa garden but, like Stourhead, there is little merit in trying to read them [7-9]. We learn many things by examining the floor plan of the villa: that the stepping of the three major *shoin* has been directed by the module of the *tatami*, for example, or that, despite the recti-linear matrix, the resulting spaces are quite fluid. But other than showing the relation of the various structures to the pond, we grasp little from the garden plan. This is a garden with a simple large idea, but with an incredibly complex development using sight, movement, and refined craft. One could generalize that this is true of most Japanese gardens. Even the plan of Ryoan-ji, which is striking as a graphic pattern, tells us little about the materials, their color-ation and the patina—and that from a seated position on the veranda it is impossible to view all of the stones at one time (although one can see them clearly in the plan) [7-10, 7-11]. The maker of these gardens would never have used a plan because plans do not capture the essence of the idea.

7-9
Katsura Villa, Kyoto, Japan,
Seventeenth century. Site plan.

7-10
Ryoan-ji, Kyoto, Japan, c. 1500.
Plan of garden with partial plan
of temple.

7-11
Ryoan-ji, Kyoto, Japan.
Marc Treib

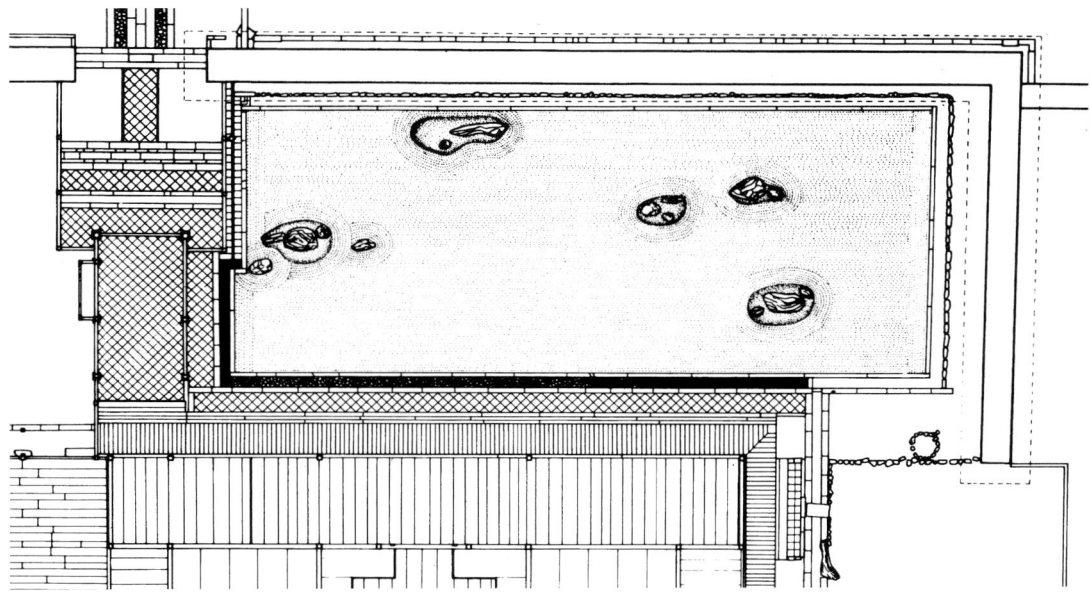

We know the plans of Vaux, of Versailles, of Chantilly and we marvel at their manipulation of geometry. We have read in all the histories of gardens about their dominant axes and how they expressed French authority and autocracy. We view on their plans the axes, the grand canals, the cross-canals, the *pièces d'eau*, the patterns of bosquets with their structures, and the architecture within the garden. But even here the plans are deceptive. Plans, as compressed geometric planar representations, disturb if not destroy our sense of the third dimension. Yes, we can identify the main axis and canals at Versailles, but how do we project the experience of walking from bosquet to bosquet beneath a canopy of the trees, or the play of animated waters? The magnificent plan of Chantilly, where expansive square meters of water appear to dissolve the land, tells us nothing of that gradual rise of land toward the statue of the Grand Condé and the sudden revelation of the water garden. Thus we see that in the great gardens, even in the most symmetrical and geometric schemes by Le Nôtre, the quality of their spaces surpass the mere layout, and the richness of our experience far exceeds the clarity of the plan.[5]

It is far more difficult to assemble a pantheon of great garden plans than one of great building plans. First, as I have suggested above, many gardens are developed without plans and are the result of gardening in space and on the land rather than designing on paper: the noun results from the verb. Second, given the broad extent of gardens, complete plans require rendering at such small scale, there is little detail to be seen. Third, landscape writers use plans as illustrations far less often than architectural historians, a trend that has been exaggerated in recent years by the wholesale publication of photo and style books. But one stunning example is Garrett Eckbo's Burden garden project in Westchester County, New York, from the late 1940s. Eckbo's fascination with the work of Wassily Kandinsky is obvious. But to his credit, Eckbo never left the garden as a flat pattern with extruded shapes. He cultivated the space of his gardens with a syncopated play between the plan and the section, and the resulting landscapes were even more complex— yet undeniably coherent—than their plans. The plan was indeed the generator, but it generated a design more significant than its own two-dimensional pattern. It would seem that this should be our aspiration for every plan, and every design.

7-12

Garrett Eckbo / Eckbo, Royston and Williams. Burden Garden, Westchester, New York, 1945. Plan.

Environmental Design Archives, University of California, Berkeley

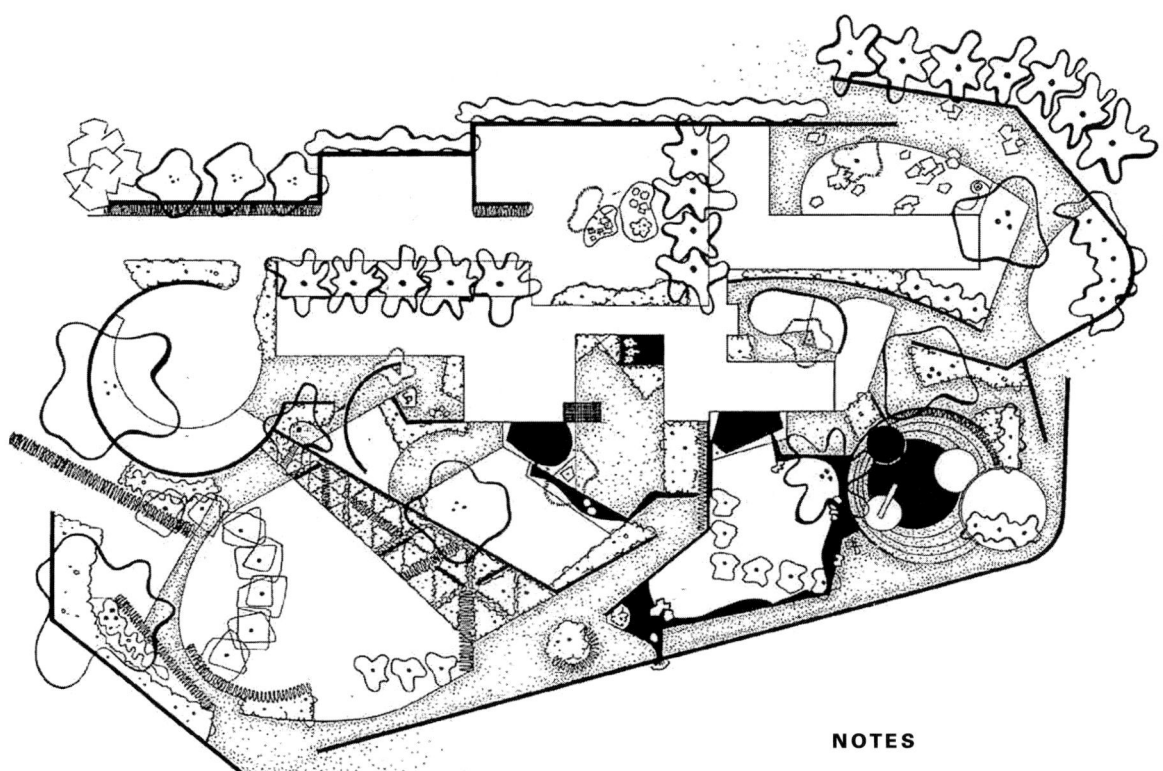

NOTES

1 Eduardo Sacriste, *Huellas de edificios /
Building Footprints*, Buenos Aires: Universitaria
de Buenos Aires, 1962. A smaller format edition
was published by the School of Design at the
North Carolina State University in Raleigh. In
1967, the Student Publication of the School of
Design published *Forty Gardens* as a companion
work. Like its predecessor, it is a wonderful
portfolio, quite revealing in showing us what
we know or don't know about these famous
gardens when seen in plan. The rendering style
lacks all the elegance of Sacriste's drawings,
however.

2 Ian McHarg, *Design with Nature*, Garden City,
NY: Doubleday, 1969.

3 Le Corbusier, *Towards a New Architecture*
(1923, reprint 1986), New York: Dover
Publications, pp. 178–198.

4 Ibid., p. 179.

5 Although we tend to be fixated on the plans
of André le Nôtre, his modeling of land form in
section is equally important.

An earlier version of this essay was published
in Swedish in *Utblick Landskap*, Number 4,
1998, as "Verkligheten från Ovan."

8

Dorothée Imbert

Skewed Realities: The Garden and the Axonometric Drawing

The connection between a system of representation and its symbolic values —between axonometry, modernity, and the experience of space—has been studied in architecture but not in landscape design.[1] Although the origins of this projection can be traced to Ancient China, landscape architects practicing in the 1920s and 1930s considered axonometry as truly modern —not only for its association with avant-garde architecture, but also for its depicting, and to a certain extent shaping of, contemporary landscape space [8-1]. Axonometry lent itself to explicating construction and organization as expressions of twentieth-century design: it simultaneously promoted the aerial view, the roof terrace, free space, and the interrelationship between indoor and outdoor. Originally focusing on the transformation of the garden, designers resorted to the axonometric view to illustrate non-symmetrical, non-axial, and non-decorative compositions. In other words, axonometry allowed them to diminish the former primacy of scenographic space and eschew any overt references to historical forms.

This elision of the recent past coincided with a rapprochement between the landscape and architecture disciplines. Landscape architects such as Garrett Eckbo, James Rose, and Dan Kiley in the United States, Christopher Tunnard in England, and Jean Canneel-Claes in Belgium, all stressed the relation between their own field and modernist architecture, whether through publications or design collaborations. And it was the domestic garden that served as their initial terrain for innovation. Reduced in size, city or suburban gardens proved an ideal field for experimentation. They were quickly built and their synthetic forms were easily comprehended, especially from that new feature of the modern house: the roof terrace. When represented axonometrically, the house and garden relationship not only appeared seamless, but spatially equal [8-2]. Although the influence of the architectural axonometric view on landscape architecture can easily be traced, others deserve mention, namely those of military treatises and art theory.

The advantages of parallel projection as a three-dimensional demonstration of functionality and constructibility have endured since the time of Leonardo da Vinci. The mechanistic and inevitable aspects of axonometric space appealed to modernist landscape architects seeking to validate their profession as both scientific and artistic. However, the potential applications of axonometry for landscape architecture and architecture differ significantly; landscape design requires little in terms of structural representation, stressing instead the interaction of ground and building. On the other hand, historical axonometric illustrations of war and battlefields offered comprehensive views of the spaces within and outside fortified walls and a clear precedent to the modern relationship of house and garden.

8-1
"Study Room under Green Phoenix Trees," Yuanmingyuan, Beijing, China, late eighteenth century.

From the album Paintings with Poems of Forty Scenes in Yuanmingyuan *(Yuanmingyuan Sishijing Tuyong)*

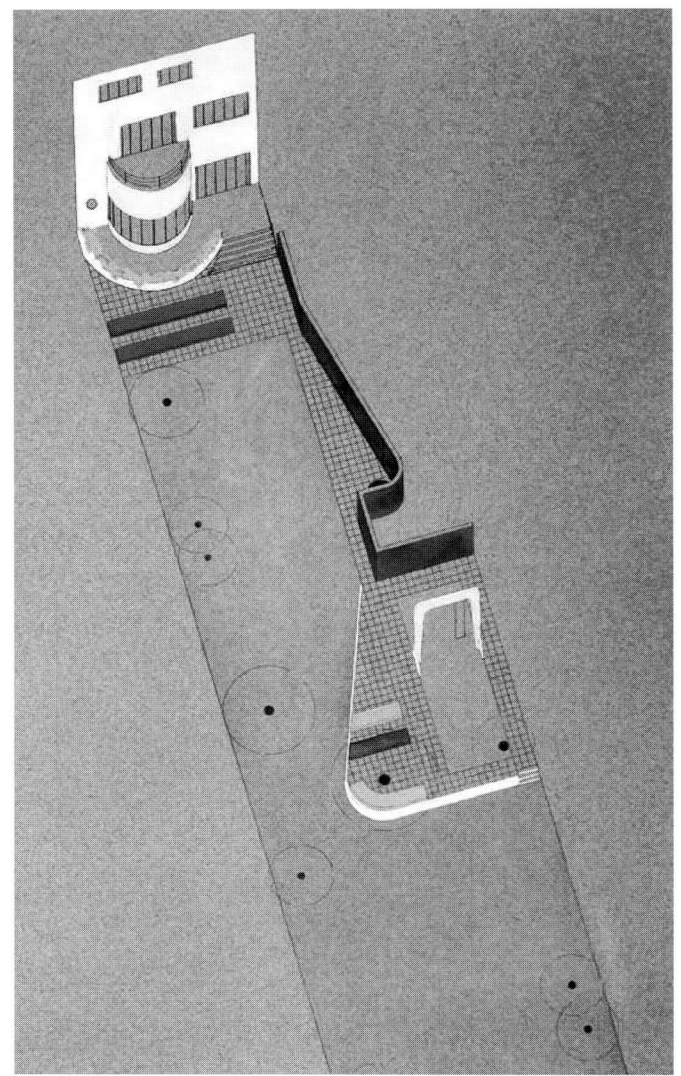

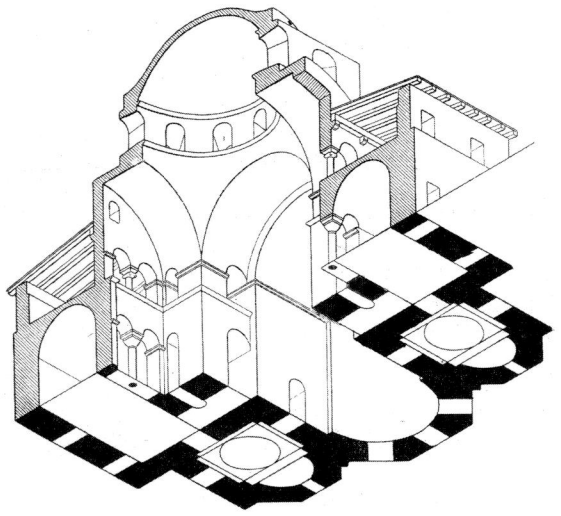

The precision of axonometry had served the ends of military architects and engineers from the sixteenth century onward by representing bastions and walls and a measurable assessment of soldiers' encampments, "seeing the entirety, distinct and clear."[2] These views were didactic and exact, technical and illustrative. Seen from above, the military terrain revealed all the elements of warfare with information developed in model form in the following century. The collection of plans-reliefs (relief maps) begun by Marshal Vauban for Louis XIV recorded in three dimensions the kingdom's fortifications, complete with townscapes and surrounding landscapes.[3] The models served not only as a tool for strategizing and improving defense, but also for publicizing the king's military accomplishments and regal superiority. The connection between the two-dimensional representation of space and form, and the plastic representation of reality—between axonometry and model as didactic and promotional tools—reappeared at the beginning of the twentieth century. In terms of modernist landscape architecture, the potential of simultaneously representing building and site underlined the connection between indoor and outdoor living.

For its technical and rational potential, axonometry later became associated with the nineteenth-century practice of descriptive geometry. Axonometry allowed precision and movement; it could simultaneously illustrate the parts and the whole suspended in the dynamic moment of assembly—all to scale. The subsequent passage of axonometry from engineering to architecture is best illustrated by Auguste Choisy's (1899) treatise *Histoire de l'architecture*, although the École des Beaux-Arts never accepted axonometry as a rendering technique for architectural projects [8-3].[4] The analytical synthesis of Choisy's plates dissected buildings and linked structure, plan, and volume into what he termed a "careful and learnedly representation of fact."[5] Expressing structure rather than perception, Choisy delineated the art of building as constructed, rather than as seen or experienced. His assemblage of inter-related elements formed a self-contained universe with little connection to the outside world—in contrast with empirical and scenographic linear perspective, which had favored the façade and its relationship to the street. Through the apparent logic of axonometric representation, efforts toward a modern landscape architecture also replaced decorative concerns with rationalism. Thus, Belgian modernist Jean Canneel-Claes employed axonometrics as a vehicle to promote a functionalist landscape during the 1930s. His was an abstract, self-defined, and complete system of interchangeable parts that supported productive and physical functions—rather than a series of spaces to be experienced frontally and sequentially.

8-2

Jean Canneel-Claes, landscape architect; Louis-Herman de Koninck, architect. Van de Putte garden, Schaerbeek, Belgium, 1932.

The steep angle of this axonometric rendering allowed Canneel to present the architectural façade as the purist backdrop to, and the inevitable conclusion of, his graphic garden.

Archives d'Architecture Moderne

8-3

Auguste Choisy, delineator. Hagia Sophia, Axonometric worm's-eye view.

Auguste Choisy, Histoire de l'architecture, *1899*

If Le Corbusier's rediscovery and later publication of Choisy's plates in the pages of the periodical *L'Esprit Nouveau* and *Vers une architecture* influenced modernist architects, and by extension landscape architects, the counter-constructions of the Dutch De Stijl movement and the writings of Theo van Doesburg also exerted a significant impact on landscape and other design disciplines [8-4].[6] As Yves-Alain Bois has stated, the rejection of perspectival and scenographic "closed space" by the Russian supremacists and De Stijl affected the representation, perception, and conception of the new architecture. The 1923 exhibition of De Stijl held at the Galerie de l'Effort Moderne in Paris truly crystallized the abstraction of axonometry. The houses, or "counter-constructions," presented at the exhibition by Theo van Doesburg and Cor van Eesteren radically challenged the representation and the very conception of architecture. The multiple layers of intersecting colored planes floating in space defied structural convention (and Choisy's scientific reality) to suggest a neutral and infinite space. Although it was unlikely that landscape architects Canneel and Tunnard had visited the Paris exhibition, their search for a styleless, rational, and elemental landscape composition testified to the influence of the Dutch movement. Their stance echoed van Doesburg's statement that plastic architecture should be "formless and yet exactly defined; that is to say, it [should not be] subject to any fixed aesthetic formal type;" it should be "economic" and "employ its elemental means as effectively and thriftily as possible;" and also that "in place of symmetry the new architecture offers a balanced relationship of unequal parts."[7] Canneel's own manifesto for the 1938 International Association of Modernist Garden Architects (AIAJM), which he disseminated with Tunnard, indirectly cited the tenets that had shaped De Stijl's counter-constructions, calling for an environment governed by geometry, asymmetry, and occult balance.[8] The text stressed that the precept of "form follows function" did not exclude aesthetic considerations as long as those were based on "harmony and rhythmic equilibrium" and that only a "concise means of expression" led to "serenity and perfect understanding of the whole."[9]

As Bruno Reichlin has pointed out in his analysis of axonometry as symbolic form, the twentieth-century version of this technique exemplified a methodological shift in architectural representation.[10] For architects like the Italian Alberto Sartoris, the axonometric view was both an illustrative construction drawing and a design tool.[11] Similarly, Canneel resorted to axonometry in diagram assembly and to test spatial relationships among landscape elements. On one hand, the accuracy in scale and precision of line suggested constructibility. On the other hand, the chromatic accents and the abstracted vegetation and topography—which only read against the architectural lines of a boundary hedge or a retaining wall—belonged to the world of pictorial representation.

8-4

Theo van Doesburg and Cor van Eesteren, *Counter-construction*.

8-5

Jean Canneel-Claes. Van de Putte garden, Schaerbeek, Belgium, 1932. Alternative design without the sports terrace, detail.

Archives d'Architecture Moderne

But Canneel also designed using axonometric methods. For example, he studied alternatives to the 1932 Van de Putte and 1933 Danhier gardens in three dimensions, inserting gymnastic equipment in the former, and examining the effect of circulation on spatial elements in the latter [8-5].

The axonometric view, devoid of all sentimentality and chance, proposed a controlled landscape to match the architectural machine. As evidenced by Canneel's manipulation of photographs to simulate mature vegetation in the publication of newly-installed designs such as the 1930 Grimar Garden, landscapes grow frustratingly slowly [8-6A, 8-6B].[12] If houses appear best immediately upon completion, it takes years, if not decades, for landscapes to attain the desired effect. With axonometry and touched-up photographs, Canneel synchronized garden with house.

The aerial view, which coopted the classical *perspective cavalière* (a military graphic construct), hinted at reality in its very detachment between airborne viewer and terrestrial space.[13] The high and oblique angle of vision of this station point avoided perspectival distortion: it paired the rationality of the plan with the added benefit of volume. The photographic representation of models achieved a similar condition, and there is a striking resemblance

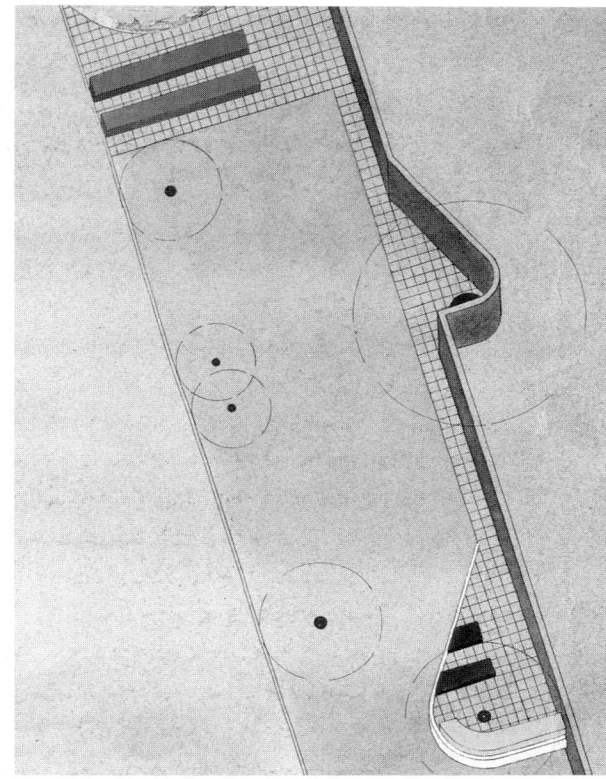

between Canneel's axonometric renderings and the oblique photograph of his Heeremans garden model.[14] If architectural modernists established similarities between photographic images of model and axonometry, landscape architects could also document the garden from a building rooftop. The trio of representation—model photograph, axonometric view, and garden photograph—depicted a somewhat synthetic space that stressed an internal composition clearly defined by a frame. This framing of the space allowed the self-referential garden to be truly modern, removing from sight any uncontrolled picturesque surroundings. Tunnard himself, even without advancing a specific formal image for the landscape, underlined the importance of the frame in the modern garden. He saw a boundary, like that of the Japanese garden fence, as exerting a steadying influence on the dynamic composition, much as the blank plane on which the axonometric drawing rested intensified the unstable equilibrium of the design.[15]

The abstract and precise technique of axonometry may seem ill-adapted for representing vegetation and topographic modeling as well as other traditional qualifiers of landscapes—tactility, texture, softness, ambiguity, and time. But the economy of means and constructed nature of axonometry appealed to Canneel, particularly during the 1930s when he exploited the narrow and elongated proportions of urban gardens to create functionalist designs [8-7]. As the oblique view favored the architectural mass in relationship to the plan or roof, it highlighted the garden as an asymmetrical organization of spaces defined by interchangeable elements and planes. Canneel achieved a singular interdependence between the space of his gardens and their graphic representation—whether crisp ink lines or flat gouaches. Such overlapping of space and painterly construct echoed De Stijl's plastic architecture.

In axonometry, garden and building not only are related, but also share equal importance [8-8]. Seen from above or below, without vanishing points, there is no far and no near. If the hierarchical and somewhat reassuring space of linear perspective places the observer in a frozen position, the spectator of the axonometric world is nowhere and everywhere at once. He or she can comprehend the entire composition with little effort, and yet cannot move within the abstracted space. Even though the axonometric view was advertised as real and rational, escaping the distortion of the perspective world, it promoted a de-anthropomorphized space, as Reichlin pointed out.[16] Alberto Sartoris recalled that he avoided altogether the favored rationalist props of airplane, street-car, or automobile [8-9]. He also refrained from including people in his drawings because he believed they would have detracted from the pure poetry of the composition.

8-6A, 8-6B

Jean Canneel-Claes, Grimar garden, Genval, Belgium, 1930. Actual state of garden and photographic montage with ink and white gouache to simulate hedges and flower beds.

Archives d'Architecture Moderne

8-7

Jean Canneel-Claes, landscape architect; Louis-Herman de Koninck, architect. Canneel garden, Auderghem, Belgium, 1931. Axonometric view with plan of house.

Archives d'Architecture Moderne

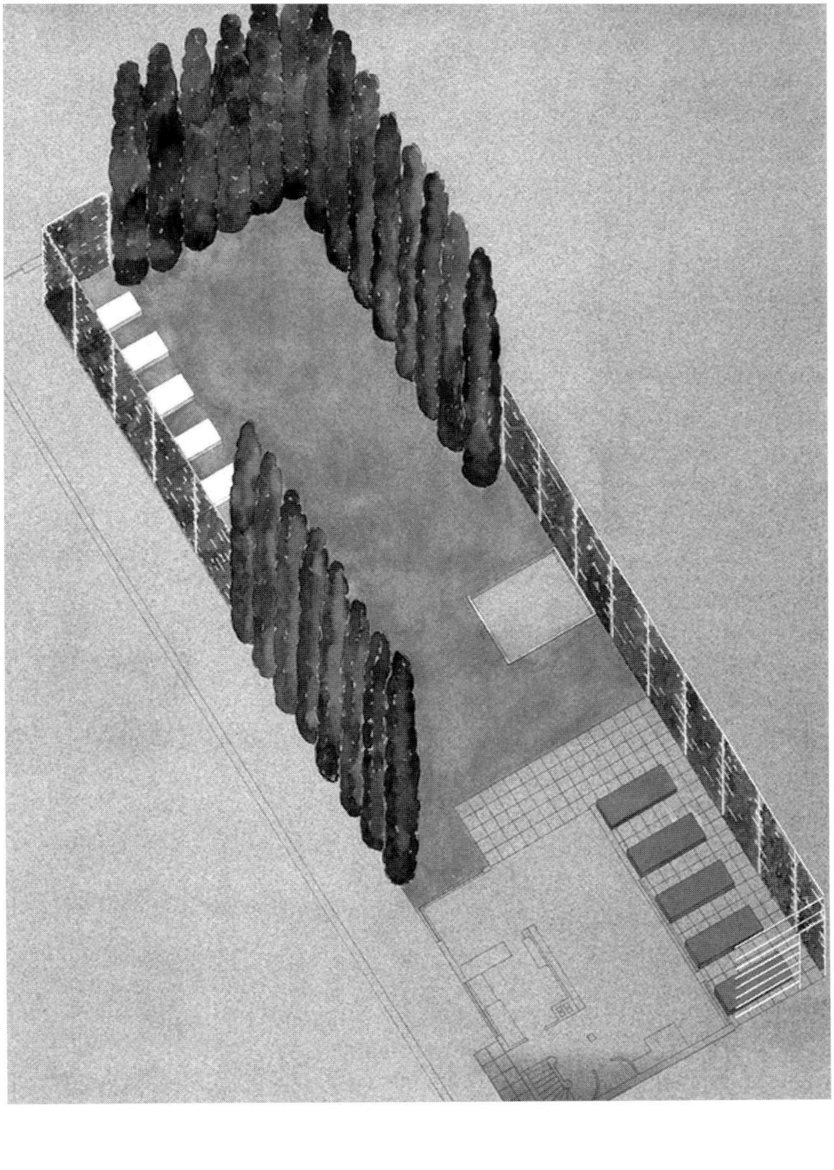

Sartoris stayed faithful to the axonometric building-as-object throughout his career, praising the efficiency of a method through which one could represent an entire project with views from two corners. Other modernist architects were less consistent in their choice of drawing. Le Corbusier used axonometry to illustrate buildings as parts of a system or a context. The illustration of the cells forming his 1922 project for *immeuble-villas* stressed the repetition and interrelation of elements within a larger machine. This seriality continued beyond the graphic border of the image to expand into the urban sphere of the Paris Plan Voisin, whose axonometric view carried overtones of scientific inevitability.[17] To depict a picturesque and dynamic reading of architecture, however, Le Corbusier resorted to perspective. Through cinematic storyboards, he illustrated an architectural promenade that moved in and out of spaces, from building to terrace, and toward the landscape. In the 1925 Villa Meyer project in Neuilly, he used both techniques. The perspective sketch expressed the sensual aspects of the architectural promenade, one which allowed for unpredictability both inside and outside. The axonometric drawing, on the other hand, depicted the structure of the villa with outdoor rooms carved from a pure volume, and as a machine immersed in the landscape.

The axonometric view is an intellectual view: it shows things as they are known to the mind and not to the eye. Canneel represented his landscapes as a series of spaces perceived from above, not as a sequence of pictures to be experienced. These images thus reinforced the modular and structural aspects of the modern garden; words could describe vegetation and texture. Canneel's axonometric landscapes did not suffer from the grandeur and vertigo effects of aerial perspective; the garden appeared compact, graphic, and utilitarian. In contrast, Tunnard illustrated his garden designs with perspective sketches by future *Townscape* author Gordon Cullen from a slightly elevated vantage point.[18] These vignettes served a dual purpose: human activity—sunbathing, strolling, even hunting—suggested livable and friendly spaces (as opposed to the machine-like systems of Canneel), while the moderately high angle of vision comfortably placed the garden within the context of the greater landscape.[19]

The pared-down volumes Canneel rendered mechanically in black ink were intended for his fellow designers, not for the greater public [8-10]. Reversible and difficult to read, axonometry expresses "a poetic viewpoint [that] calls for specialized knowledge."[20] Canneel published his formal investigations in architectural periodicals like the Belgian *Bâtir* and *La Cité* rather than in popular home and garden magazines. Like Sartoris, he proposed a product finished and perfect: no part could be modified or removed without compromising

8-8

Jean Canneel-Claes. Danhier garden. Fort-Jaco, Belgium, 1933.

The 45° angle of the axonometric view features the house as an integral furnishing of the garden-orchard. Canneel structured the site with three architectonic plateaux inserted in a matrix of trees and registering the slope.

Archives d'Architecture Moderne

8-9

Alberto Sartoris. Van Berchem house and studio, Paris, 1930.

Alberto Sartoris

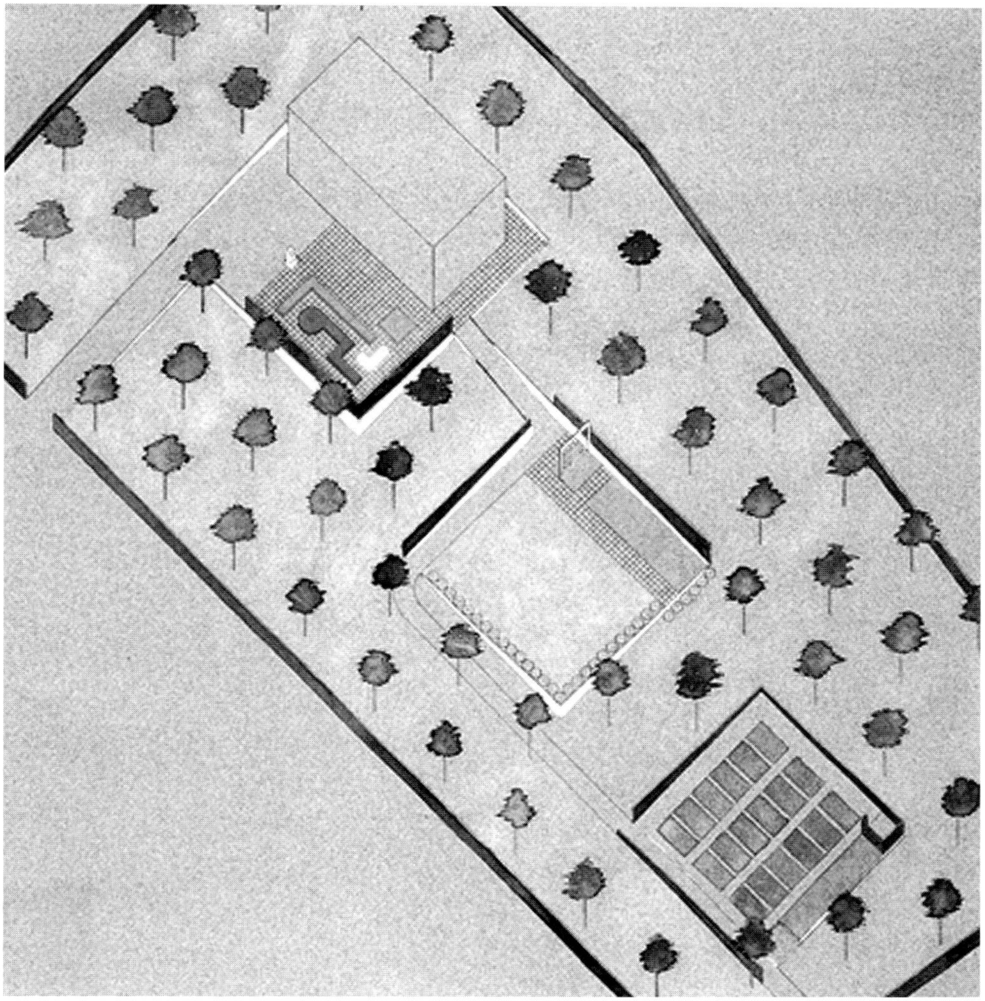

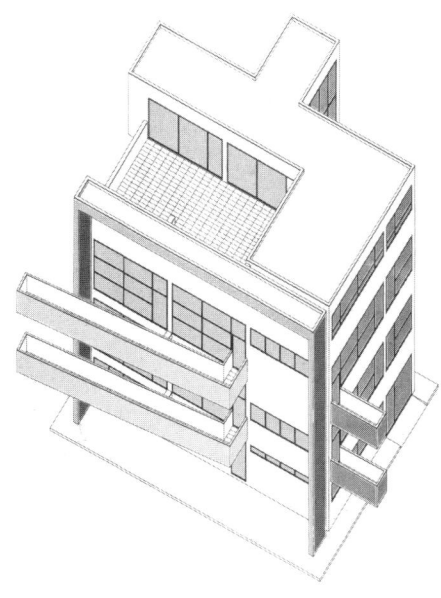

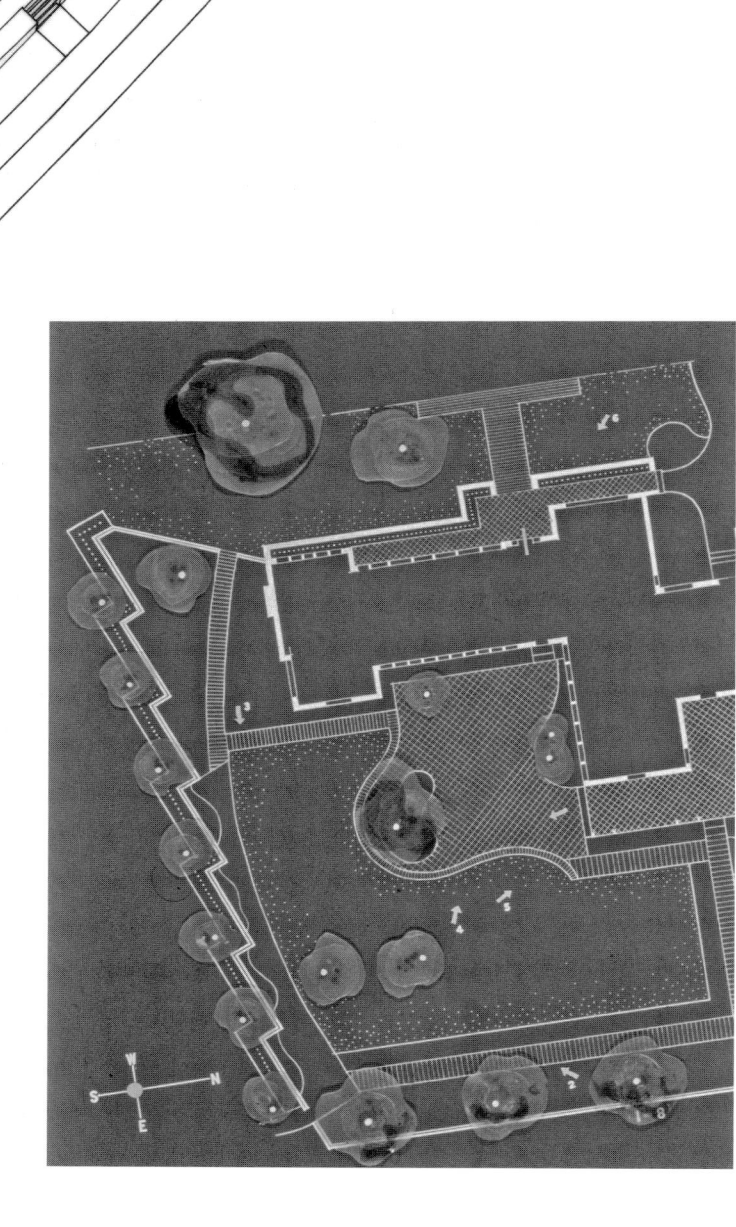

the balance of both plan and space. Axonometry brought the garden to parity with the house; it also portrayed a sculptural, painterly world in which time and movement were suspended. Canneel's understanding of the potential of axonometry as a vehicle to publicize his concept of modernity is quite evident. Simply stated, axonometry looked modern. But Canneel was not alone in his attempt to update the rather conservative field of landscape architecture through representations.[21]

With its yearning for modernity, the early twentieth century witnessed the swelling of the third dimension in landscape representation. French landscape designers André and Paul Vera featured the garden in plan and its background in axonometry, borrowing both from a classical garden representation technique and a modern vocabulary that reinforced the flatness of the ground plane and the sense of enclosure. Gabriel Guevrekian juxtaposed plan, elevation, and volume in the painterly representation of his 1925 Garden of Water and Light in Paris and underscored the complementarity between the Heim house and garden terraces with the axonometric gouache. Thomas Church—or more accurately, Robert Royston and Serge Chermayeff, who worked for Church in the late 1930s—marketed modernity in the garden with plans from which celluloid trees protruded [8-11]. This unusual combination of plan view and model elements, and of white lines on a brown field also advertised modern gardens as functional and artistic. The boards were exhibited in Cargoes, a San Francisco store where enlightened patrons (mostly architects) could view and order furniture by Alvar and Aino Aalto, and possibly a modern garden to go with it. By combining the highly graphic nature of its design with a suggestion of reality, the garden was advertised as a product.

Similarly to Canneel, Garrett Eckbo, Dan Kiley, and James Rose borrowed from the axonometric view from architecture in their crusade for a modernist landscape, publishing their theoretical stances in *Architectural Record* and *Pencil Points*.[22] In this way they could stake out their territory before a viable audience, while educating that audience about the validity of modernist landscape architecture. Eckbo's "Small Gardens in the City" of 1937 stressed formal relationships within the gardens as well as adjacencies within the urban block, both in axonometric and model forms [8-12].[23] Although Eckbo resorted to axonometry to stress the modular variations of gardens within an urban system, he seemed reluctant to surrender phenomenological concerns. The sun shines on his gardens, casting shadows and canceling the gravitational and temporal neutrality of the axonometric representation as people stiffly inhabit the space. It was as if Eckbo had chosen the architectural medium of axonometry to reinforce the systematic and volumetric aspect

8-10
Jean Canneel-Claes. Buzon garden. Schaerbeek, Belgium, 1929.

The highly abstract ink drawing appeared in the architecture and urbanism periodical *La Cité* in December 1931, with other early projects of Canneel.

8-11
Thomas Church. Hiatt garden, Modesto, California, c. 1938.

Environmental Design Archives, University of California, Berkeley

of his gardens, and yet could not abandon the perceptual aspects of the landscape. Thus he introduced his designs with this thought:

> Gardens are places in which people live out of doors. Gardens must be the homes of delight, of gaiety, of fantasy; bold or free arrangements of space and material will generate such feelings and responses. Designs shall be three-dimensional. People live in volumes, not in planes.[24]

If axonometry best suited the representation of Canneel's simple graphic products, it did not always survive Eckbo's complex and texturally-rich world. Nonetheless, Eckbo and Rose also exploited the potential of the axonometric projection as a working tool and a diagram for assembly, with elemental compositions that avoided fixed aesthetic types [8-13]. Rose described an exploded axonometric whose "modular parts…organized separately… demonstrate their interlocking relation. The parts are fitted together with each part and open space indispensable to the other." Axonometric views which, surprisingly, Rose derived from scale models and plans, emphasized "the range and flexibility of a modular system." In these vignettes, viewed slightly higher than eye-level, characters interacted socially—shaking hands, raising a hat—to confirm that the rational system of planting, paving, and fencing fostered a human space. In an exploded axonometric diagram, hedge,

8-12

Garrett Eckbo. Small Gardens in the City (project). San Francisco, 1937.

Eckbo countered the abstraction of the axonometric composition by including sun and shadows, and people and textures.

Environmental Design Archives, University of California, Berkeley

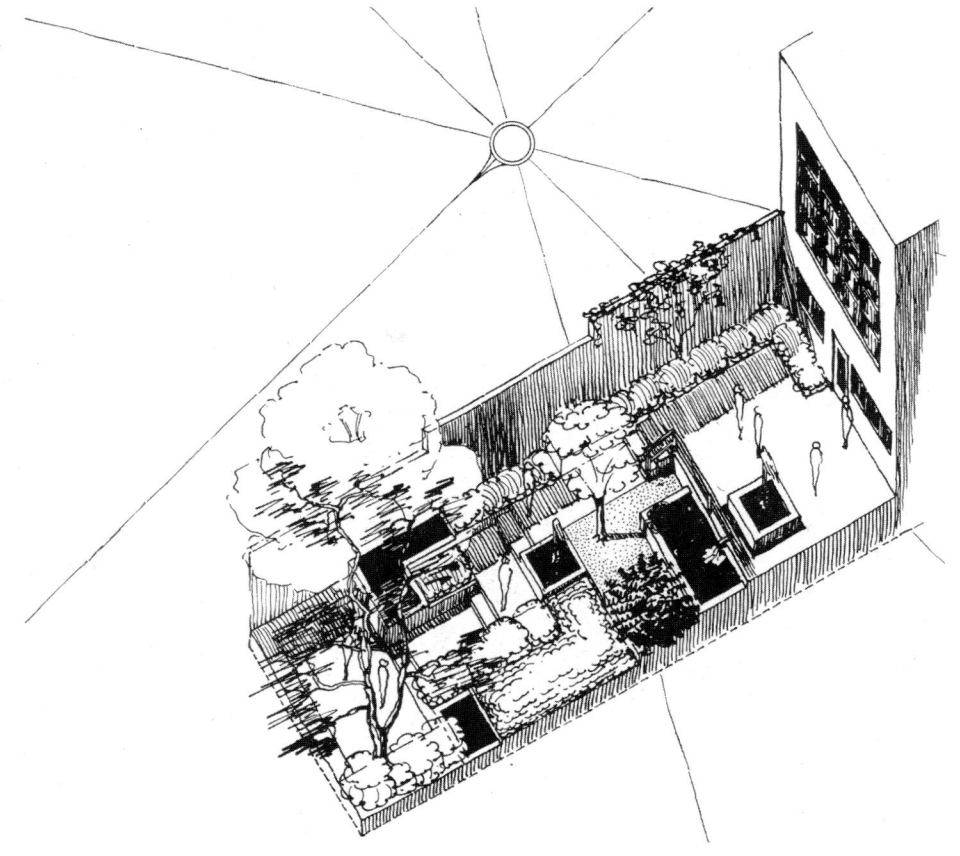

trellis, pool, and tiles, formed a kit of parts to be assembled into a modern outdoor volume, functional and dynamic. Eckbo's and Rose's axonometric investigations expressed no perfect reality like that of Canneel's floating gardens. Using the medium as a tool by which to study proportions and complicated volumetric relationships—rather than as a rendering or advertising techniques—they often produced diagrams of design concepts that were less than clear. Axonometric visions such as Eckbo's (1941) Burden garden, or Rose's own garden in Ridgewood, New Jersey, composed overlapping planes, intersecting angles, and overhead canopies to reinforce the abstruse perception of the axonometric view.[25] The ease with which any uneducated eye can grasp the empirical space of linear perspective disappears in the axonometric layers. Perspective is drawn for the consumer, axonometry for the producer—that is, for the architect and landscape architect.[26]

Within the architectural realm, axonometry weakened the sense of frontality and the importance of the façade, as the ground or ceiling plane came to dominate. In the garden, this shifting from the human body as datum to the bird's-eye view confirmed the supremacy of the plan, making it an integral part of the spatial composition. The ground plane could be perceived as a play of geometries, colors, and textures not only within the drawn world of

8-13
James Rose. Pool Garden for the *Ladies Home Journal*, 1946.

In this project, Rose sought to demonstrate the value and flexibility of a modular system (3' x 3') for garden design.

James Rose, Creative Gardens

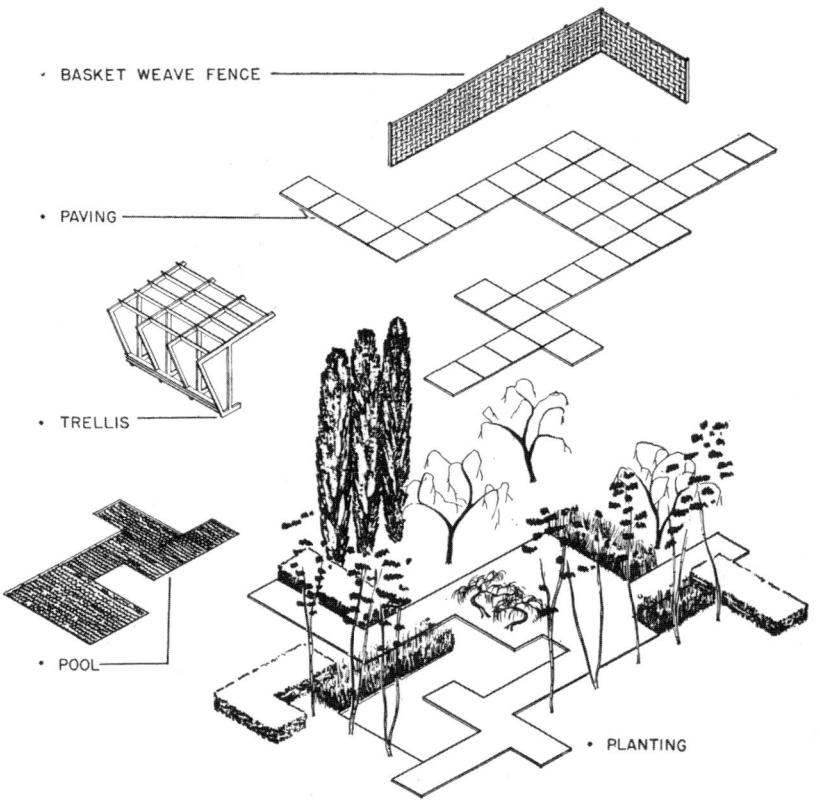

· BASKET WEAVE FENCE

· PAVING

· TRELLIS

· POOL

· PLANTING

Canneel but also in the garden seen from above. It is ironic perhaps that axonometry, with its anti-perspectival denaturalization and dissolving of context, would become a favored tool of representation for modernist landscape architects.

In conclusion, one could advance that axonometry not only conferred a sense of modernity on landscape representation, but also shaped a certain type of space. Axonometry allowed Canneel, Eckbo, and Rose to assemble various functional and plastic elements into ideal combinations. But Canneel remained singular in expressing a clear relationship between thought and product. To him, the axonometric garden synthesized the representation of modernist space and a design process; the technique also lent itself splendidly to his minimalist compositions. Inert and living materials held the same value, with architecture and landscape architecture sharing equal footing. The oblique angle favored asymmetry and the balance of unequal parts. Axonometry detached the spatial composition from its environmental matrix and at the same time suggested that garden and house formed but one element of a greater system. The house and garden became an *organigramme* (the French term for a synthetic structure diagram), suggesting a variety of uses with minimal means and echoing Reyner Banham's description of the Choisy view as an "immediately comprehensible diagram."[27]

As if they were designing in a sandbox, this cadre of modernist landscape architects employed a limited palette firmly bounded. The axonometric view upheld the idea of the modernist garden as abstract, synthetic, and asymmetrical, but it also implied a garden-object more tightly connected to architecture than to the greater context. As landscape architects relaxed their dependence on architectural theory and moved from garden design as a vehicle for experimentation towards systems of increased scale and complexity, they abandoned axonometry and its perfect, self-contained, world.

NOTES

1 As Yves-Alain Bois and Massimo Scolari have pointed out, parallel projection has followed a long and mostly undocumented trajectory. For a discussion of the sources of, and attitudes towards, axonometry, see Bruno Reichlin, "L'Assonometria come progetto," *Lotus International*, 22, 1979: 82–93; Yve-Alain Bois, "Metamorphosis of Axonometry," *Daidalos*, 1(1), September 1981: 41–58; Massimo Scolari, "Elements for a History of Axonometry," *Architectural Design*, 55(5-6), 1985: 73–78. To put it simply, axonometry is an abstract representation of a three-dimensional object onto a plane, without convergence and thus with an equal distribution of detail, and to scale. Axonometric projection comprises isometric, dimetric, and trimetric views. For a description of orthogonal projection and a brief history of axonometry, see Jean Aubert, *Axonométrie: Théorie, art et pratique des perspectives parallèles*, Paris: Éditions de la Villette, 1996. There is a certain amount of confusion regarding whether isometric axonometry distorts the plan, or not. Architects like Alberto Sartoris favored orthogonal isometric axonometry, or military axonometry, for its true plan. In other isometric representations, such as Herbert Bayer's 1923 drawing of Walter Gropius's office at the Weimar Bauhaus, the two axes of the plan are set at 120 .

2 Jean Aubert and Massimo Scolari both cited the Venetian military engineer Giovan Battista Delicci, who in 1538 described parallel projection for military constructions as being "true." See Aubert, *Axonométrie*, p. 84. See also Bois's account of the application of axonometry to military art and nineteenth-century technical drawing in "Metamorphosis of Axonometry," pp. 51–56.

3 The making of scale models of fortified towns for military purposes can be traced back to a representation of the city of Rhodes in 1521. Louvois, Minister of War for Louis XIV, commissioned the first plan-relief from Vauban in 1668 for the town of Dunkirk. See Isabelle Warmoes, *Le Musée des plans-reliefs: Maquettes historiques et villes fortifiées*. Paris: Éditions du Patrimoine, 1997.

4 Choisy learned descriptive geometry from Jules Maillard de la Gournerie at the École Polytechnique. *De la Gournerie's Traité de géométrie descriptive* (1873–1885) included "*perspectives axonométriques et cavalières.*" If the theory and practice of axonometry appeared in German and English publications from the mid-nineteenth century onward, it was not the case in France where the representational

schism between engineers and architects endured. Architects used descriptive geometry for shaded and shadow-casting details, and conical perspective to represent buildings and space. Apparently, perspectival representation retained its stronghold on architectural design at the Beaux-Arts until the school's architectural department closed in 1968. See Aubert, *Axonométrie*, pp. 85–88.

5 Choisy, cited by Reyner Banham in *Theory and Design in the First Machine Age*, London: The Architectural Press, 1960, p. 25.

6 The contrasting modernity of Roman, Byzantine, and Gothic architecture abstracted into worm's-eye views would appeal to Le Corbusier, who himself paired Greek architecture with automobiles. Le Corbusier resorted to the plan-biased representations of Choisy, such as Hagia Sophia, to stress how plans generated the entire structure of buildings. See Le Corbusier, *Vers une architecture*, Paris: Éditions Crès, 1923, pp. 36–38. Tunnard emulated Le Corbusier, whom he cited on numerous occasions when discussing the tyranny of styles and the importance of the plan in garden design. Even Tunnard's chapter entitled "Towards a New Technique" was a reference to Frederick Etchells' 1927 translation of *Vers une architecture* into *Towards a New Architecture*. See Christopher Tunnard, *Gardens in the Modern Landscape*, London: The Architectural Press, 1938, pp. 69–73.

7 See "1924 Theo Van Doesburg: Towards a Plastic Architecture," in Ulrich Conrads (ed.), *Programs and Manifestoes on 20th-Century Architecture*, Cambridge, MA: MIT Press, 1971, pp. 78–80. In fact, the interplay of dissimilar elements achieving a rhythmic composition held a strong appeal for several modern landscape designers. In 1930, Fletcher Steele praised Pierre-Émile Legrain's 1924 garden in La Celle-Saint-Cloud for its "occult unsymmetrical balance" and the dynamic relationship of its parts. See Fletcher Steele, "New Pioneering in Garden Design," *Landscape Architecture*, 20(3), April 1930: 163–164, 172, 177.

8 See the manifesto for the Association Internationale des Architectes Jardinistes Modernistes (AIAJM). Copies can be found in the papers of Belgian architect, Huib Hoste, at the Katholieke Universiteit Leuven and in the Tunnard papers, 1938–1956 file, at the Landscape Institute in London. Point 5 of the manifesto states that garden design should respond to context and functional demands with flexibility and asymmetry. Tunnard quoted almost verbatim two points of the manifesto's English text, crediting neither the association nor Canneel.

Tunnard, "The Garden in the Modern Landscape," *Architectural Review*, March 1938: 131.

9 See points 2 and 6, AIAJM manifesto (French text).

10 Referring to a renewal of representation, Reichlin described axonometry as far more "explicit and concrete than a diagram but sufficiently abstract" not to be confused with a model to be reproduced. Axonometry serves as a sort of "metadesign," situated at the intersection of functional and artistic needs and the structure that will satisfy them. See Reichlin, "L'Assonometria come progetto," p. 86.

11 Ibid., p. 86.

12 See the "doctored" prints of the Grimar Garden in the Archives d'Architecture Moderne in Brussels.

13 The expression *perspective cavalière* apparently derives from the sixteenth-century military representation of the terrain observed from a cavalier, a term for an earthen promontory affording a sweeping view over the fortifications and their surroundings. Another interpretation of this projection in which parallel lines remain parallel refers to the view of a horseback rider (also a cavalier) onto an object below.

14 There are other photographs of the Heeremans model that do not illustrate the relationship between house, garden, and slope as clearly as the view published in Christopher Tunnard's *Gardens in the Modern Landscape*, p. 80. See Hoste papers, Sint-Lukasarchief. On the connection among axonometry, architectural model, and photography, see Gérard Monnier, "Perspective axonométrique et rapport au réel," *Techniques et Architecture*, February–March 1985: 122.

15 Tunnard, *Gardens in the Modern Landscape*, p. 84.

16 Györgi Lukacs, *Estetica*, Torino: Einaudi, 1973, vol. 1, pp. 52–53, cited by Reichlin, "L'Assonometria come progetto," pp. 88, 92.

17 In the film *L'Architecture d'aujourd'hui* (Pierre Chenal, 1930–1931; music by Albert Jeanneret; 13 minutes), Le Corbusier dramatically inserted one of his towers in the transformed Parisian fabric of the Plan Voisin.

18 Gordon Cullen, *Townscape*. London: The Architectural Press, 1961.

19 Tunnard, *Gardens in the Modern Landscape*, pp. 72–76.

20 See Bruno Reichlin, "Reflections: Interrelations between Concept, Representation and Built Architecture," *Daidalos*, 1(1), September 1981: 67.

21 Historically, scenographic depictions of gardens (whether real or imaginary) were created for enlightened patrons to satisfy their Cartesian or pastoral yearnings. But the transcription from mind to paper to ground usually left few tracks. If documentation was essential for constructing elaborate hydraulic machinery and delicate fountains, the idea of the garden appeared to glide effortlessly from simple drawing to tapis vert and bosquet, from the mind of Le Nôtre to field hands.

22 See, for instance, James Rose, "Freedom in the Garden" and "Why Not Try Science?" respectively in *Pencil Points*, October 1938 and December 1939; and Garrett Eckbo, Daniel U. Kiley, and James C. Rose, "Landscape Design in the Urban Environment," "Landscape Design in the Rural Environment," and "Landscape Design in the Primeval Environment," in *Architectural Record*, May 1939, August 1939, and February 1940.

23 Garrett Eckbo, "Small Gardens in the City," *Pencil Points*, September 1937: 573–586.

24 See "Small Gardens in the City," and Garrett Eckbo's commentary on this text in "Pilgrim's Progress," in Marc Treib (ed.), *Modern Landscape Architecture: A Critical Review*, Cambridge, MA: MIT Press, 1993, p. 209.

25 For Rose's modular experiments and garden axonometrics, see his *Creative Gardens*, New York: Reinhold, 1958. For the axonometric and model views of Eckbo's Burden garden and other projects, see his *Landscape for Living*, New York: Duell, Sloan & Pearce, 1950, pp. 162–163, and Marc Treib and Dorothée Imbert, *Garrett Eckbo: Modern Landscapes for Living*, Berkeley, CA: University of California, 1997.

26 Reichlin, "L'Assonometria come progetto," p. 86.

27 According to Reyner Banham, *Theory and Design in the First Machine Age*, pp. 24–25: "The formula is: isometric in its setting out, it presents plan, section and elevation in a single image, detailing is suppressed and one is left with an elegant and immediately comprehensible diagram."

9

Laurie Olin

Drawings at Work:
Working Drawings,
Construction Documents

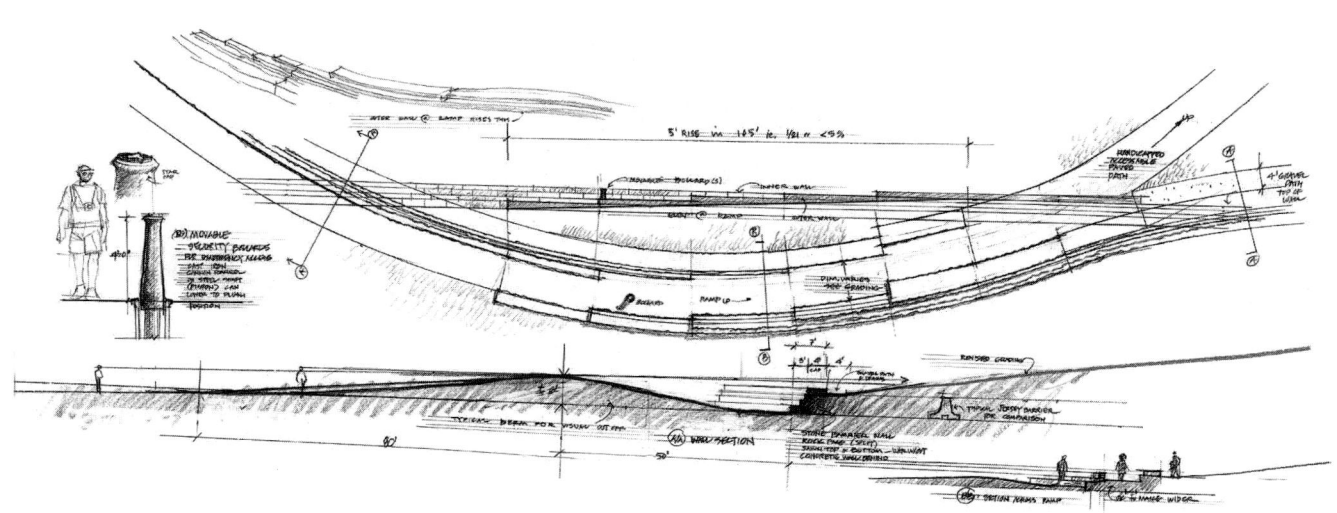

Paul Ricoeur once remarked that work is the essence of what people do, and that it ranges from manual labor to contemplation. Drawing is the work of designers. Whether it is done with a computer or a pencil, drawings are what we actually make. No matter how much we think of ourselves as builders and makers, landscape architects like myself almost never actually build or make anything physical ourselves. Someone else does, and almost always from our drawings.

What today are called "construction documents" (CDs) used to be called "working drawings," and many of us still use that term. I prefer the older name even though it avoids the eventual legal situation implied by the current, official title. This name, of course, derives from the simple fact that the drawings are provided to the people who give physical form to our projects: the masons, plumbers, carpenters, nurserymen, bulldozer operators, laborers, surveyors, and layout engineers. Our work, these drawings—the products of our hands, eyes, and imaginations—are really only the instructions upon which others base their work, which is the product of their hands, their eyes, and their understanding. Those of us who try to design things and get them built—as opposed to those who commission, plan, or study them—frequently complain that things would be fine if only contractors would follow our drawings and specifications. It is also true that contractors are forever pointing out the inadequacies and lack of clarity—or worse, confusion—in our drawings. Thus, construction drawings represent different things to different people. To further complicate matters, there are at least two generations of construction drawings for complex projects: the first being the designer's documents and the second consisting of shop drawings. These are something one doesn't hear much about in school, but which have everything to do with what arrives on the job site. Shop drawings are provided by manufacturers and suppliers such as stone quarries and metal fabricators, or as in a recent project in our office: glass manufacturers and millwork shops.

So, what then are construction documents? How are they conceived and produced? What are the issues involved? How are they different from other design drawings? How have they changed, and how are they evolving at this moment? Some of these issues will receive more attention than others, but I offer them for consideration as questions of genuine interest and concern to the field and those who wish to build.

Design students, regardless of field, rarely understand that the drawings they are learning to make are only the beginning of a process, and that design doesn't stop when the construction documents begin. In fact, design doesn't stop after construction documents are complete, but continues

The painter sketches to paint, the sculptor draws to carve, the architect draws to build.

Louis Kahn, 26 August 1962

9-1
Olin Partnership.
Washington Monument Grounds, Washington, DC. Iterative study in plan (above) and cross-section (below) of the ha-ha earth grading, path, and granite security walls.

Olin Partnership

throughout the construction period with a myriad decisions— even those made daily during the heat of construction. At times these difficulties are exacerbated in realizing landscape architecture, partly because the subject matter—unlike a teapot or a building—is notoriously difficult to describe. The evolution of a design and the further resolution that takes place during construction may be as minor as determining a connection or fitting, or positioning a joint between two elements. On the other hand, it can be more serious, such as resolving a conflict between two uncoordinated trades —most commonly between structural and mechanical features (an absolutely ubiquitous and seemingly eternal problem known to all practitioners) and the effect upon the integrity, appearance, or workability of the scheme.

Sometimes, however, there are major intentional design changes in the field, or at the final hour between the completion of the documents and building. Only the construction drawings record these modifications. Technical or financial concerns often cause these changes, but the instigation can also derive from new insight gained upon reviewing the work under construction. While working on the landscape for Bryant Park behind the New York Public Library, I noticed that a mid-block drive and a series of pylons drawn by the architects Carrère and Hastings around 1910 had never been executed. I strongly suspect that cost factors had also forced several portions of the front façade to remain unfinished for several years after the opening of the building, as is evident in early photographs. The lions that flank the library's front steps on Fifth Avenue—the lions which have become such an icon of New York City and a symbol of the library itself—were not shown in their current location on the construction drawings, but instead were originally intended for the smaller 42nd Street entrance. Why? Who knows? This was probably just one of those last minute changes believed to have been made for the better. Construction documents are thus both devices for communication and negotiation. At times designers put more into a project and the documents than they need or expect to get, thereby building in a reserve of items which can be sacrificed for cost savings or to be used for negotiation at a later time during the long and often arduous period of bidding and construction.

Some drawings are made to record information, to select and point out particular aspects of the work, to help us see something. Some are created to persuade, to present an argument, or to entice—as in presentation drawings. Some attempt to figure out a problem, to support the search for solutions or yearning for a particular arrangement or character. Some drawings are created to consider ideas or to communicate general intentions or particular concepts. Construction documents, however, are created solely for the purpose of explaining how to make things. For this reason there are properties found in other forms of drawing not commonly present in CDs, namely, expression, feeling, and mood. Conversely, there are several nearly universally shared conventions that allow people of all sorts to make, read, and understand a wide range of construction drawings. Many of these conventions have been in use for centuries, some even for thousands of years.

There are examples of construction documents made with ink on papyrus dating to the eighth dynasty of ancient Egypt. That they were intended for construction is instantly recognizable to anyone familiar with the conventions of measured drawings and orthographic projection—two of the most fundamental devices underlying all design drawings. In one example exhibited at the National Gallery of Art several years ago, the scale and some indication of the profiles and depths of some of the elements were missing. Yet with the materials and joinery, the finished effect, relationship of the parts to each other, and the overall proportions and detail, the drawings were quite clear.[1] We can assume that these features were conveyed either by other documents or through verbal and drawn instructions on the job site and what today are called addenda, or Requests for Information (RFIs).

While most of the instructions for constructing landscapes and buildings during classical antiquity and the Middle Ages have been lost, a good record remains from the Italian Renaissance onward. The most fundamental drawing is the plan, a measured drawing that indicates layout, dimensions, elements, and in some cases materials and details of construction [9-1]. The elevation is a graphic convention for depicting the vertical and horizontal relationships of things on façades, or at other elements set 90° to the horizontal plane of the plan. In addition to these are sectional views: cuts through an object or the land. Most designers today are probably so familiar with these fundamental orthographic projections that it is difficult for them to consider them as arbitrary and invented. With a number of other conventions promulgated between the sixteenth and twentieth centuries regarding the use of fixed proportional relationships, or scales, a hierarchy of line weights, drawing orientation, presentation sequence, symbols, and schedules, these drawings constitute the basis of construction drawings today. This period of over four hundred years has witnessed significant changes in the methods and technology of production, the number of drawings needed, and the means of construction upon which they address. Despite this evolution, an experienced designer today can nearly always understand the information presented in contemporary construction drawings by a foreign professional or those of an earlier era.

Probably the most imaginative of all these conventions is the section, for it is truly the most conceptual of all drawing types. Renaissance artists introduced the concept of seeing through surfaces to an underlying structure, an X-ray vision of sorts that has proved enormously useful in understanding and describing how things work. A literal example of such a view would be a drawing of an orange sliced perpendicularly to its axis, revealing the triangular wedges of its segments within the round circle of its outer skin. Among the best early examples of sections are Leonardo da Vinci's anatomy studies and the cutaway views of the structure of St. Peter's Basilica in Rome prepared by Baldassare Peruzzi to explain his concept to the Pope.

If asked to draw an existing or proposed garden, few of those without a landscape architectural education could create an adequate plan or elevation, let alone a section drawing as nearly all such drawings are highly conceptual. They represent hypothetical slices or cuts made horizontally or vertically through walls, floors, ceilings, trees, streams, streets, utilities, structures, and earth with all manner of things that cannot really be seen or experienced simultaneously. Three devices make such drawings possible: (1) the Euclidian concept of a three-dimensional grid (with its x, y, and z axes) in which any point or object may be positioned in space; (2) standardized measurement; and (3) the notion of scale, whereby things can be represented and visualized at a significantly different size than they are in reality.

While equal in age to drawings for building, landscape documents have experienced greater difficulty in achieving accurate representations. A significant problem for landscape is that of portraying multidimensional and plastic ground surfaces, namely, topography. One traditional method first creates a recognizable view and then alters it by redrawing or overlay, annotating the new view with comments on the work intended. A study for Claremont made by William Kent shortly before his death in 1746, for example, gives instructions on how to shape the land in a vista from the house [9-2].[2] It is a sketch view supported by some few words in a fluid scrawl and a good example of an early attempt to address this problem of topographic alteration. A generation later, Humphry Repton illustrated his business card with a picture showing the landscape gardener using a level on a tripod, while in the distance workers reshape the terrain. The image certifies that Repton is adept at using the recently developed technique of land surveying with scientific instruments.[3] Despite these advancements in method, Repton nonetheless continued to utilize the graphic methods used by Kent, but developed them into a sophisticated method of overlay presentations of key views. These were bound together and presented in what became known as the Red Books (see Stephen Daniel's chapter in this volume for examples).

9-2

William Kent. Claremont, Surrey, England. Sketch for landscape adjustments, c. 1746.

British Museum

9-3

Pennsylvania Avenue, Washington, DC. Sketch proposal for grading, dimensions, and planting by Thomas Munroe, sent to Thomas Jefferson, 1803.

Library of Congress

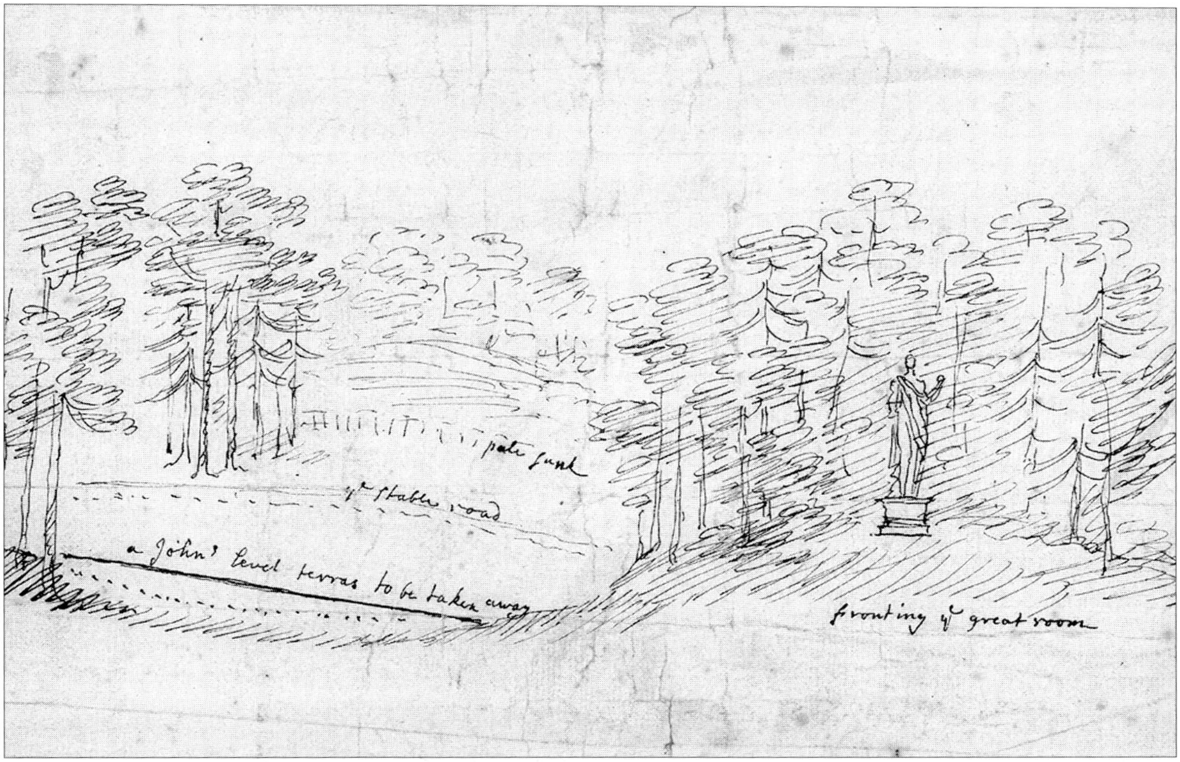

pale sunk

y^e stable road

a John! level terras to be taken away

fronting y^e great room

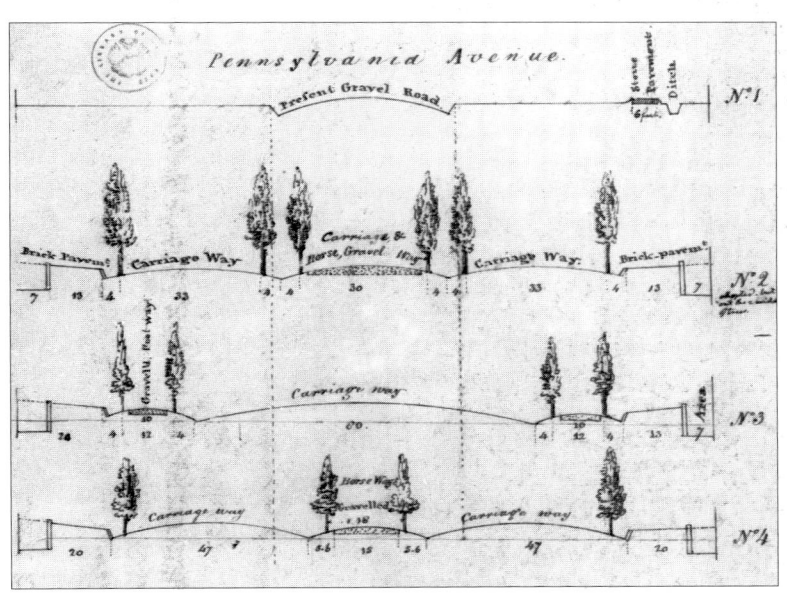

Landscape depictions of grading techniques explored a variety of methods up to and through most of the career of Frederick Law Olmsted, Sr., relying on hachure, spot elevations, illustrative views, and various azimuths with percentages of slope from particular points.

Landscape construction documents of the generation between Kent and Repton—while clearly indicating where to plant particular trees, how to shape water bodies, or make particular features such as bridges, ha-has, and follies—remained almost totally silent about site grading. For centuries, direct field supervision, rather than instructions in drawn form, prevailed. On-the-spot direction—waving the arms, staking, and exhortation on the part of the designer, a clerk of the works, or an engaged owner/client—best conveyed how to shape the land in three dimensions. The exception to this generality was the occasional use of the cross-section to prescribe the modeling of roadbeds and other regular landscape forms. An excellent cross-section from the period appears in the correspondence between Thomas Munroe and Thomas Jefferson regarding the proposed dimensions and grading of Pennsylvania Avenue in Washington, DC [9-3].[4]

A breakthrough in early nineteenth-century France became widespread by mid-century: the use of contour lines to depict topography. The new convention was rapidly adopted by the municipal bureaus that rebuilt much of Paris later in the century and significantly aided their work. Proudly published by the director, Adolphe Alphand, in *Les Promenades de Paris* are engravings after the original engineering drawings that record the grading for the Parc des Buttes-Chaumont, built between 1863 and 1867. The original drawings for the existing and proposed topography were drafted with steel nib pens in red and black ink—not a particularly common practice at the time (but easily achievable today with our computer-controlled ink-jet printing).[5] The resulting images were both stunning and highly effective.

Historically, the siting of buildings, the construction of landscape structures, and the location of trees may have been fairly precise, but for centuries the prescriptive documentation for shrubs, herbs, perennials, and annuals has been much looser, more personal, idiosyncratic, vague, and highly prone to subjective interpretation. Plans prepared in 1749 by Richard Woods for an herbaceous border at Philip Southcote's renowned *ferme ornée*, Woburn Farm, and those by Thomas Jefferson for flower borders at Monticello, consist of the written names of plant species distributed on a plan. This drawing type evolved with garden styles into the convention of using annotated blobs and shapes to indicate the masses and beds of similar plants, especially shrubs and flowers as they came to occupy a major element of nineteenth-

9-4

Beatrix Farrand. Garden for Mrs. Herbert L. Satterlee, Bar Harbor, Maine. Planting Plan (detail).

Here Farrand adopts the technique developed by Gertrude Jekyll for planting in "drifts" for her herbaceous borders.

Environmental Design Archives, University of California, Berkeley

and early twentieth-century designs.[6] Some of these planting plans—including those made by Gertrude Jekyll, Beatrix Farrand, and Ellen Biddle Shipman —are elaborate [9-4]. Even so, much was left to be determined in the field by the nurserymen and gardeners. On the whole, however, little changed in landscape construction documents in the century and a half between Kent and Alphand except for the adoption of contour lines and the introduction of photographs, pioneered by the Olmsted firm and subsequently by individuals such as Gertrude Jekyll.[7]

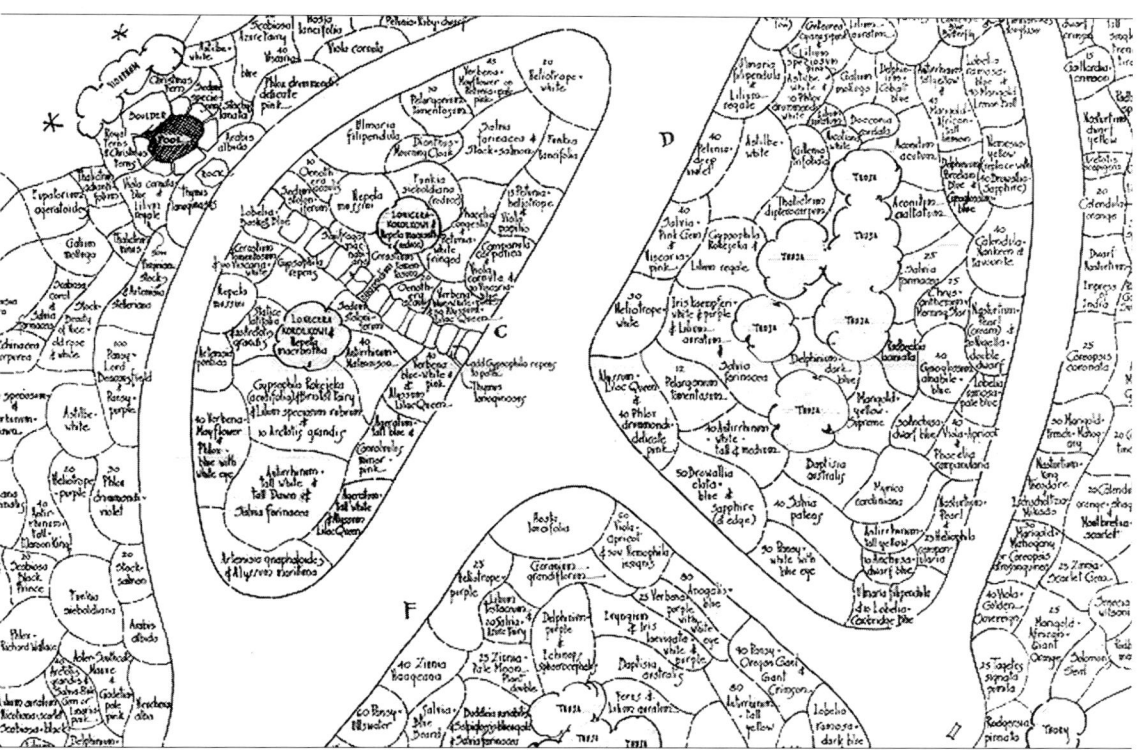

Standardized conventions regarding the organization and general format for drawings and the order of work emerged in the twentieth century as architecture and landscape architecture became more specialized, more professional, more homogenous, and more regulated by both the state and professional organizations. Today's sets of construction documents normally follow this sequence: a survey of the existing site; site and erosion protection, demolition, areas designated for equipment and the storage of supplies, arboriculture and any other special site preparations; layout; materials and paving; grading and drainage; construction details of site features; and planting. The drawings prepared by other professionals follow, structural and civil engineering first, then mechanical, plumbing, electrical, lighting, irrigation, and other specialties such as graphics design, security, and communications.

So many assumptions and abstract devices are employed that drawings—although seemingly self-evident to seasoned practitioners—appear to laymen as mysterious and unintelligible marks sent from another planet or prepared in code by some secret society. Consider the following three examples. One shows a portion of a layout drawing with its radii, curve data, and clouds, which indicate a change in the information from the original bidding set [9-5]. Another might be the common abbreviated note "op hand sim," which means that there is a mirrored symmetry of dimensions and materials, i.e. "the opposite hand (on the left, as opposed to the right, or vice versa) is similar or symmetrical" to the detail drawn. A drawing from one of our office's early projects shows the challenge of conveying an accurate picture of the materials and grading [9-6]. The drawing indicates the specific layout and jointing of a set of stone steps set within a field of herringbone-patterned brick paving. The steps adjoin a wall that contains a series of trees, each surrounded by a particular tree grate. The most abstruse symbols here are the contour lines indicating the shape of the earth to be graded behind the wall. While any professional worth his or her salt is familiar with this method of representing topography—and some like myself admit to a great fondness for grading—these lines are essentially phantoms, a total fiction. Contours are a marvelous device for describing vertical and horizontal relationships, the invention of which solved a centuries-old problem in landscape design and construction documentation. Similar to a ring around a bathtub or any shoreline, they indicate a series of imaginary levels (used to describe vertical form), practitioners must understand their principle, imagine their existence, and transform them into action. But unlike the lines depicting the edges of bricks, or the changes in plane of steps, they do not exist in reality.

While certain processes such as grading can be effectively represented by plans, some require other forms of explanation. Sheets of wall details, for example, still normally contain the now-ancient cross-section. Our drawing for a stairway on a hill indicates the concept in a very direct way, also showing the desired method of incorporating a glacial erratic boulder into a retaining wall and a landing of the stair required by code [9-7]. Flicking through almost any set of drawings used to solicit bids reveals such issues. In addition, schedules of materials allow contractors to place orders with their suppliers. If the designer cares about how it all goes together and what it will look like—or how well something will withstand wear and tear—a detailed paving schedule will supplement the drawings. Numerous sheets of this sort are needed for any project with significant amounts of masonry, for example, our projects for the 16th Street Transit/way in Denver, the National Gallery of Art Sculpture Garden, in Washington, DC, and Bryant Park [9-8]. For work of any sophistication, every dimension, joint, shape, and condition must be studied, understood—and drawn. Failing to do so will always lead to unpleasant consequences.

Once a builder or subcontractor places an order for the production of custom items indicated on the drawings—whether for benches, millwork, metalwork, stone paving, treads, or copings—the supplier prepares shop drawings to demonstrate their understanding of the specified items. These drawings are then sent to the designer for approval—an extremely critical phase of the work requiring great patience and an eagle eye. Any error or misunderstanding between the stone cutters or the welders and the landscape architect will almost certainly create havoc on the job site.

In the case of planting, documentation remains chronically imperfect. Oddly enough, in an attempt to be precise, construction drawings almost never exhibit any of the quality and character of the plants they document. In this they differ radically from design studies that impart the feeling of particular plants or convey the final planting effect. CDs may actually succeed in achieving the effect depicted in the former, but they normally look (and feel) hard-boiled and vapid, or at times even feeble by contrast—as do many of the plants and trees when first installed. As everyone familiar with landscape knows, they'll grow up and out and fill in, hopefully to match the richness of earlier design studies and presentation renderings.

Trees, as suggested above, are relatively easy to position in construction drawings; not so the numerous smaller plants we use. Plant schedules are usually quite clear, giving exact quantities, species, varieties, and sizes with notes regarding special installation requirements. Detail drawings indicate the planting depth and pit preparation for every plant, as well as the desired spacing for areas of mass planting. On the other hand, just where and exactly how they are to be planted inevitably ends up being determined in the field. For example, to avoid unnecessary effort, wasted drawing, and to help with site supervision, we recently plotted large drifts of mixed grasses and native perennials for a four-acre rooftop meadow in Salt Lake City using blobs and attenuated shapes resembling peanuts and amoebas. But, despite these specifications, my partner Susan Weiler decided the actual mixing, spacing, and arrangement on site. In the past, our office has used grid systems for similar situations, but we found that they needed as much, if not more, adjustment in the field as the "blob" method. In fact, once installation begins, loose drawings and general shapes drawn onto the site are really all that is needed. On project after project, no matter how precise or thoughtful the drawings, one still needs to turn and adjust trees and shrubs to their best

9-5

Olin Partnership. Toledo Museum of Art, Toledo, Ohio, 2000. Construction plan detail.

The "clouds" indicate revisions to the original layout instructions for stonework and walkway.

Olin Partnership

9-6

Hanna/Olin, Ltd. Johnson and Johnson Headquarters, New Brunswick, New Jersey, 1978. Construction plan drawing indicating detail arrangement and layout of paving, steps, trees, walls, and grading.

Olin Partnership

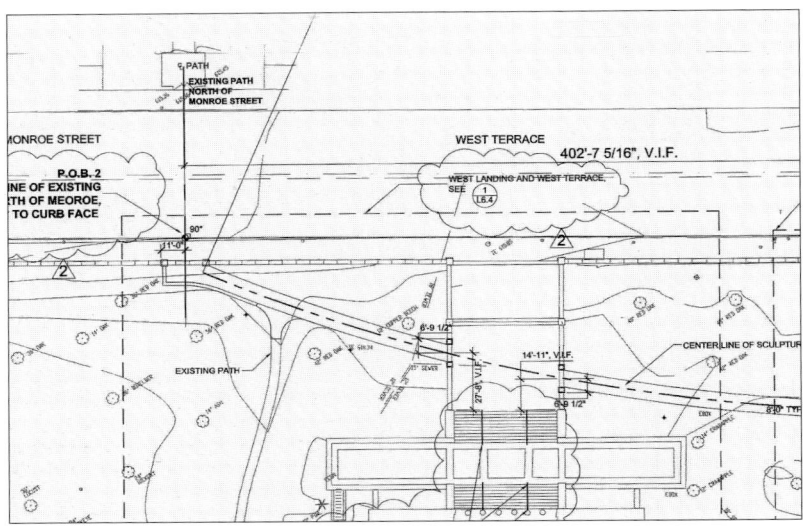

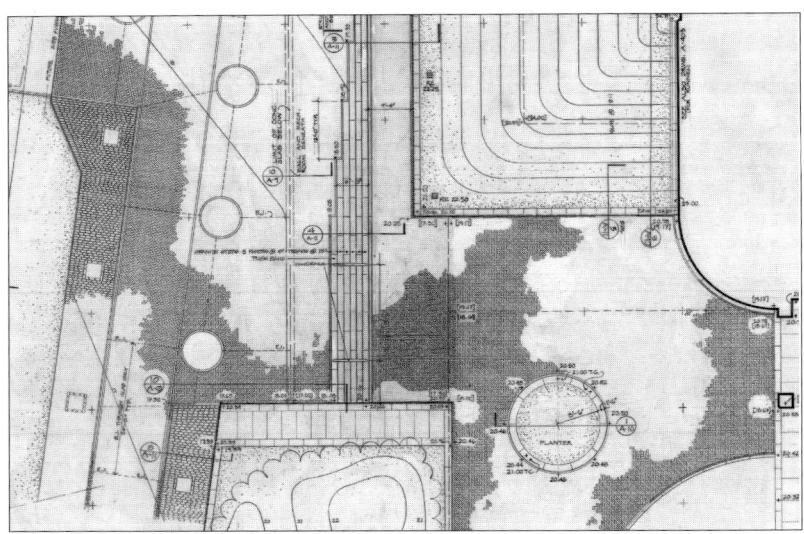

9-7

Hanna / Olin, Ltd. Pitney-Bowes Headquarters, Stanford, CT, 1981. Construction document of cross-section showing the relationship desired between a proposed stair and a glacial boulder from the site.

Olin Partnership

9-8

Hanna/Olin, Ltd. Bryant Park renovation, New York, 1988.

Stone paving of the upper terrace demanded detailed paving documents depicting every "dimension, joint, shape, and condition."

Olin Partnership

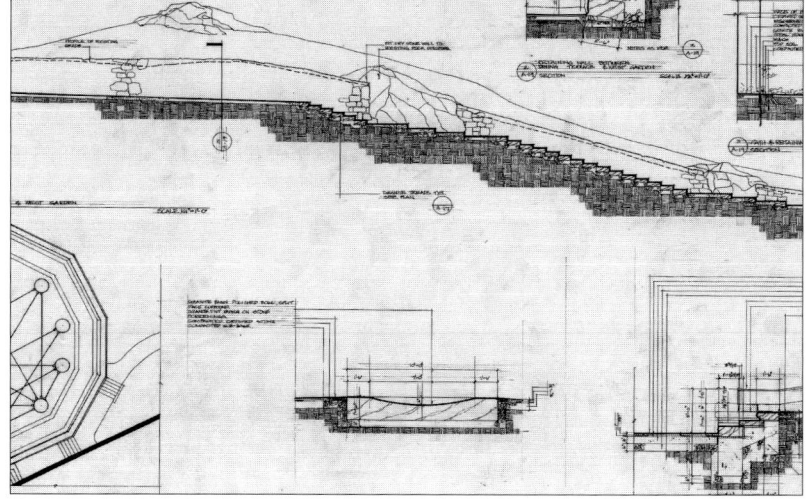

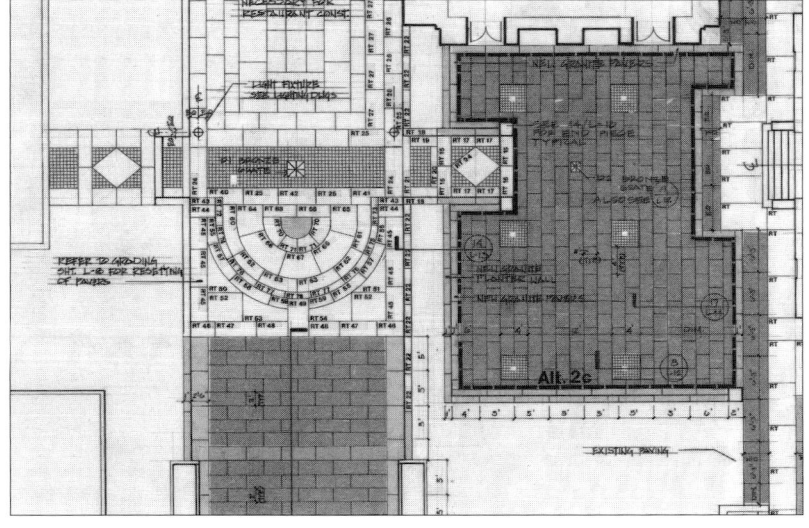

advantage on site. In the case of the cactus garden at the Getty Center in Los Angeles, my partner Dennis McGlade had to put the gloves on and move the fuzzy little devils around until they looked and felt just right.

Although we have occasionally used modes of representation such as models and photography to explain our intentions—even incorporating them into the construction documents, I think it is fair to say that drawing remains the most common method for documenting landscape construction. There have been continual experiments and changes in methods through history and even in my own lifetime these have been rapid and dramatic. In 1956, when I first worked for the Alaska Road Commission, architectural and engineering drafting (still spelled "draughting" then) used ink with ruling pens on sheets of coated linen cloth. For lettering, one used either small steel Crowquill pens dipped in bottles of ink or Leroy stencil guides traced with a drafting pen. The linens were then reproduced via a process that exposed them upon sensitized sheets of paper thereafter put through a chemical solution in huge vats and hung out to dry—this produced blueprints with white lines. This was a slow and cumbersome process that required enormous patience. By the time I left architectural school in 1961, however, the ammonia-based diazo system that produces positive blueline and blackline prints and sepia reproducibles had come to dominate the field.

Professionals then drew with pencils on sheets of heavy vellum paper. Old timers who had worked with Wright, Aalto, and Mies sharpened their pencils with razor blades, licked the leads, and wiped them on their shirts. Photostatic prints and Rapidograph drafting pens became popular in the 1960s. By the early 1970s wax pencil leads on plastic mylar film had become widespread; full-size film negatives made in vacuum frames, then spotted and masked with opaquing paint, signaled the adoption of photographic processes. These negatives were printed to create positive copies. Cutting sheets of Zipatone or Pantone patterns or color-tinted film and rubbing it onto slick Chronopakes (photo-reproductions) for presentations or construction documents were common for several decades. Lettering evolved from rub-on transfer letters to a contraption called a Kroy machine, now as extinct as the slide rule or the early giant Marchant calculators. The production of construction documents as layered sheets of information on mylar, all overlaid and registered with punch bar strips taped to the top of drafting tables, anticipated the current practice of computer-aided drafting (CAD). Today nearly every student coming out of undergraduate and graduate landscape architectural programs around the world is conversant with CAD, Photoshop, Illustrator, Form*Z, Rhino, and a myriad of other computer programs. We send drawing files electronically between offices across the country and around the world, sharing base drawings and standards, printing out the latest changes made by our professional consultants and collaborators wherever they or we may be.

Like others in the profession, our office is caught up in this evolution, trying to turn it to our purposes, and with increasing success. On the plus side, the steady development of documents from "schematic design" through what used to be called "design development" (now often called "50-" or "60% construction drawings") proceeds on to final "bidding documents" as a continuous sequence. We no longer have to throw a complete set of drawings away only to start over after each phase of the work is completed. This is a major improvement.

Unfortunately, significant drawbacks plague the use of state-of-the-art CAD systems. The vast areas depicted in many landscape drawings do not fit very comfortably on small monitors, which rarely exceed 21 inches in width. Even buildings, which are almost always smaller in plan area and fit better within the limits of the screens, are seen only in fragmentary views—and at no particular scale. If the entire landscape document is visible, it is too small to read, much less be evocative of feeling. Conversely, if you zoom in (or is it out?) to a scale sufficiently large to comprehend any particular segment, its context and relationship to the whole disappear. This might be cured by a screen the size of an old-fashioned drawing board. But while the Pentagon and the Massachusetts Institute of Technology may already possess such devices, I fear that it will be some time before landscape architects possess the means to purchase screens of these dimensions, should they come to commercial market.

On the plus side, one can print out enormous, even full-size, drawings on today's plotters. On the negative side, half-baked, poorly thought out, tentative first drafts—or even unworkable solutions—may appear finished and glowingly resolved when viewed on the screen. CAD simulations also convey a disembodied lack of material density and appear luminous but numinous. It's also very expensive and just as tiring, labor-intensive, and eye-straining as old-time pen and ink drafting, and getting the work out of the machines, "plotting," can be remarkably time-consuming. The refrain that the network, or server, or the plotter "is down" has become familiar to everyone in the field.

Nevertheless, today we all use this technology to produce construction drawings. For all of our recent projects, for example, that of the new Stata Center for the Computer Sciences at the Massachusetts Institute of Technology, a landscape design our firm developed with Frank Gehry's office in Los Angeles, the final construction documents were prepared on computers.

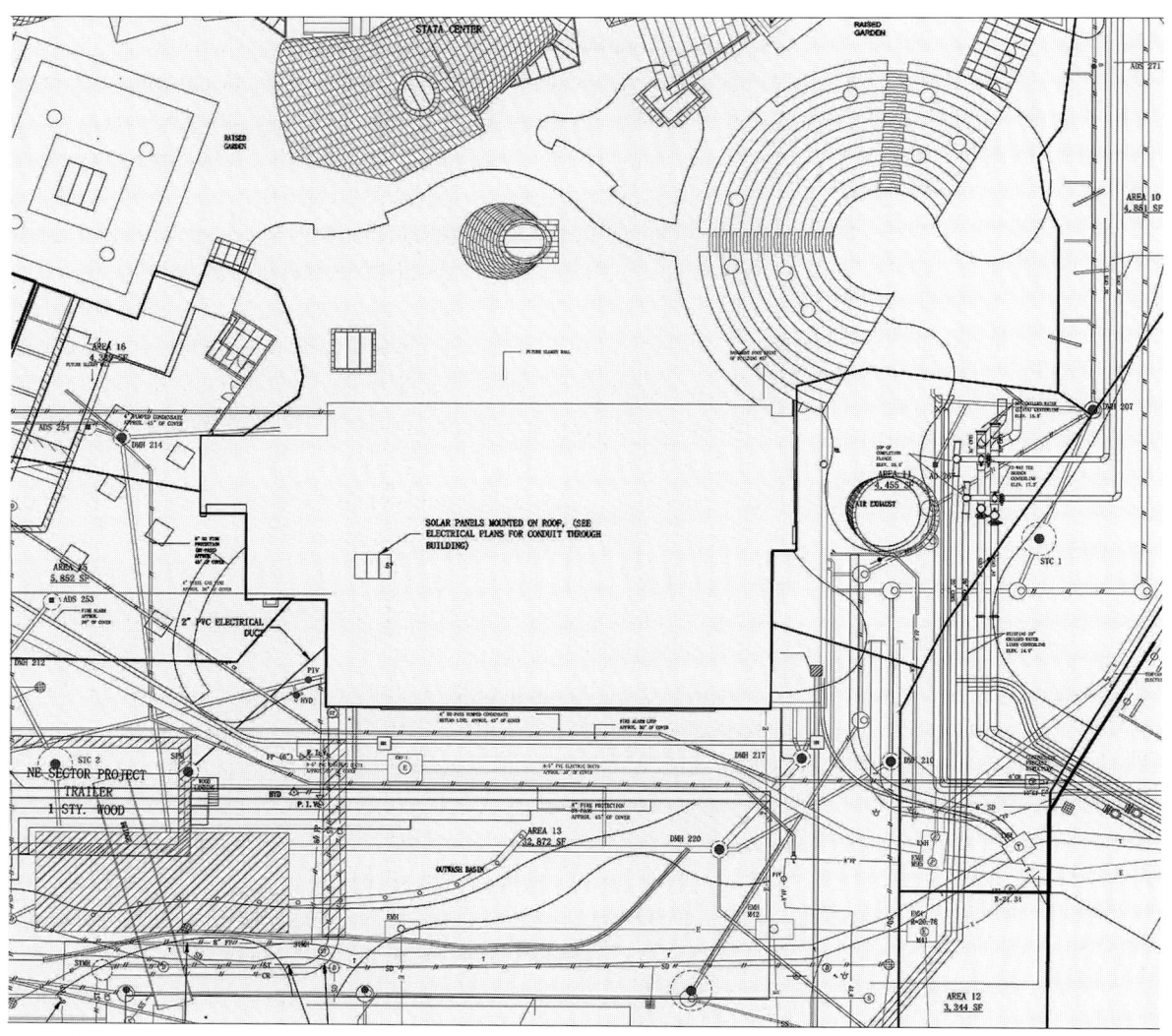

For us, I had to admit reluctantly a few years ago, CAD technology for CDs is a near miracle, and not just due to the ease of communication and coordination with the architects (which is significant) but also for its ability to coordinate subsurface mechanical systems with our landscape work. The illustration [9-9] shows a portion of our Landscape/Utilities coordination drawing pertaining to a fairly sophisticated storm water management system that we developed with the engineers of Judith Nitsche in Boston. Consisting of various infiltration basins, scrubbing, polishing, and detention elements, all overlaid with other structures and trades seen here in something resembling an X-ray. Other benefits are in the offing. Gehry's office, for example, has begun sending architectural and engineering files directly to manufacturers. From these, structural members or cladding can be cut directly from the designer's construction drawings using computer-controlled machinery, thus eliminating the difficult and costly step of shop drawings.

Even so, it is our experience that the use of computers in the production of construction drawings must occur as an iterative process paired with drawings and sketches done using more traditional techniques. Drawing is an act of thinking. If you can't draw something, you probably can't make it. Additionally, free-hand studies are still one of the most powerful, convenient, and useful design tools ever devised.

There are times when there is no better way to conceive and direct particular aspects of landscape construction than with free-hand sketches. The design and realization of a fountain at the Getty Center provide a case study. After recording in a sketchbook a set of suitable rocks, I sketched out a possible composition on the flight returning from the field. While I suppose we could have translated these studies into another format, there didn't seem much point, as the sketches explained the idea quite well [9-10A, 9-10B]. Using the sketches alone, we constructed a full-size mock-up in the woods at Columbia, near Yosemite National Park, where we had found the stones. After adjusting them in the field, we took some Polaroid photos, made a few additional sketches with dimensions, marked the stones, trucked them to Los Angeles, and recreated the arrangement in the Getty Center Museum Courtyard. In this case, the construction drawing was also the concept and design drawing.

Finally, I must admit that I am almost always interested in seeing construction documents created by designers I admire or whose work I find interesting. Rarely published and scattered about, these drawings offer an insight into the creation of particular projects, as well as the resourcefulness, craft, and concerns of the designers. Consider the layout drawing for Lawrence Halprin's innovative Lovejoy Fountain and plaza in Portland, Oregon. It shows

9-9
Olin Partnership. Stata Center for the Computer Sciences, Massachusetts Institute of Technology, Cambridge, MA, 2001.

A portion of the Landscape / Utilities coordination plan suggesting the complexity of infrastructure today and the need for clarity and precision in the construction drawings that direct them.

Olin Partnership

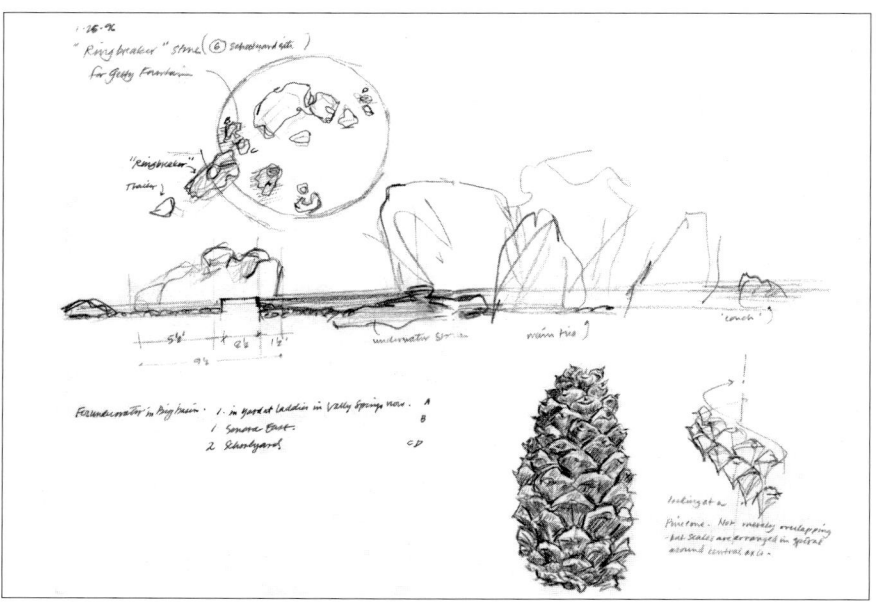

the clear influence of the chipboard contour models constructed during its design. It is drawn in ink. While the intentions are clear, much of the actual desired geometry resides in the drawing, not in numbers, calculations, or elaborate survey set-ups. The fountain portion is enlarged in greater detail and described with numerous cross-sections and dimensions [9-11]. Nonetheless, much is left to be resolved in the field by the builder and the landscape architect's field representative. The climate of construction in America has changed so much in recent years that few offices today could issue a document such as this with any assurance that it would be followed successfully.[8] More recently, I have found that, like my own office, Halprin and his staff are in the habit of drawing and mocking things up full-size before committing them to final construction documents.

What isn't in these drawings? When I first visited Halprin's work in Portland I was struck by the high quality of the concrete work, the attractive aggregate finish and the fine handling of formwork and finishing. I wondered how much of this was due to the specifications and how much was the result of superb field supervision from Halprin's office, or a fabulous contractor and crew. What was the formula for the concrete mix, and where was the aggregate from? I don't think scrutiny of the drawings would answer these questions.

In contemporary practice, a companion document accompanies every set of construction drawings: the specifications. In earlier times, notes on the

9-10A, 9-10B
Laurie Olin. Getty Center, Los Angeles, California, 1995. Sketchbook studies of the courtyard fountain.

Olin Partnership

drawings comprised the verbal remarks necessary to communicate procedures, finishes, manufacturers' standards, approvals, and disclaimers. Today, shifts in technology, materials, and manufacturing—with changes in the training and skills of construction workers and the corporate industrialization of both the construction and design industries—have forced change in the form of the specifications. This is also partly due to the increase in legislation and litigation governing safety and liability. In response, construction documents have become far more detailed, exhaustively explanatory, and specific. So, too, has the writing become vastly more detailed, technical and extensive —often consisting of multiple volumes for large and complex projects. As a comment on this situation let me add that in the thirty years of the practice of Hanna/Olin and the Olin Partnership—with the exception of a failure of one light pole and one pump (both designed by sub-consultants)—all our sorrows regarding lawsuits derive from issues arising from non-compliance with the procedures and methods specified—namely, from matters documented in the specifications—and not from any of our drawings. It is not surprising to us as landscape architects that these troubles have had to do with matters of soil or underlay of pavement. Nearly every problem that has caused us to seek legal assistance regarding a construction project has been due to something that can't be drawn. One conclusion, therefore, is that if you can draw it, you can probably build it. Another is: that the better you draw it, the greater chance that it will be built better—and possibly even in the way you intend.

ACKNOWLEDGMENTS

I would like to thank Marc Treib for asking me to reflect on this topic, which initially I didn't think would be much fun. I have never taught construction, per se, and despite many years of practice and teaching—probably like most people in the field—I had not reflected much about this important and not well-understood or often discussed, but absolutely central topic in our field. In the course of finding the material and thinking about it, I realized how much of my life I had either been making them myself or looking over others making them. I love construction drawings, for they are how we get things done. There is nothing quite like a beautiful set of working drawings—except, of course, a beautifully built place.

NOTES

1 This drawing was included in the exhibition "The Renaissance from Brunelleschi to Michelangelo: The Representation of Architecture" that began at the Palazzo Grazzi and was later shown at the National Gallery of Art in Washington, DC, in 1994. It is reproduced in the catalog of the same title edited by Henry A. Millon and Vittorio Magnano Rizzoli; Lampugnani, 1994, p. 20.

2 The sketch referred to is one of several made for Claremont now in the British Museum. With others, it is discussed in John Dixon Hunt, *William Kent, Landscape Garden Designer: An Assessment and Catalogue of his Designs*, London: Zwemmer Ltd., 1987.

3 Reproduced in Stephen Daniels, *Humphry Repton: Landscape Gardening and the Geography of Georgian England*, New Haven, CT: Yale University Press, 1999, p. 11.

4 Munroe was in Washington trying to get aspects of the capitol built. His letter with these alternatives was sent to Jefferson on March 14 1803. Jefferson received it at Monticello on the 18th and responded on the 21st with a choice (he picked no. 2) and sent along a cross-sectional sketch of his own to insure that there would be no mistake. Munroe's alternatives and Jefferson's response are now in the Library of Congress and are reproduced in Frederick Doveton Nichols and Ralph Griswold, *Thomas Jefferson, Landscape Architect*, Charlottesville, VA: University of Virginia Press, 1978, p. 68, and in Frederick Gutheim, *The Federal City: Plans and Realities*, Washington, DC: Smithsonian Institution Press, 1976, p. 141, respectively.

5 The original folio of *Les Promenades de Paris* published in Paris in 1873 can be found in most university or architecture and design libraries. In 1984, Princeton Architectural Press, New York, issued a reduced format facsimile. The plate referred to is located in a section of unnumbered illustrations toward the back in the section entitled "Les Promenade Intérieures de Paris," and is itself labeled "Plan des Courbes de Niveau du Parc des Buttes Chaumont."

6 A discussion of Wood's planting, and this and other documents of his, may be found in Mark Laird, T*he Flowering of the Landscape Garden: English Pleasure Grounds, 1720–1800*, Philadelphia: University of Pennsylvania Press, 1999, p. 103 ff. Jefferson's planting plan(s) for

portions of Monticello may be seen in Nichols and Griswold, Thomas Jefferson, fig. 48, and ff.

7 For Jekyll's use of photographs, see Judith Tankard and Michael Van Valkenburgh, *Gertrude Jekyll: A Vision of Garden and Wood*, New York: Harry Abrams, 1988. Although no study has appeared to date on this particular aspect of the Olmsted firm, it is clear from the voluminous amount of materials published during the past thirty years that their office was devoted to the use of photographs, considering the enormous amount of pictures taken on site before, during and after construction of one project after another. Beyond a desire to document and record their prodigious achievements, there is clearly a utilitarian motive behind many of these photographs, used in the office for reference. These still reside in the Olmsted office, Fairstead, in Brookline, Massachusetts. Good examples can be seen in Elizabeth Barlow [Rogers] and William Alex, *Frederick Law Olmsted's New York*, New York: Praeger, 1972, and Cynthia Zaitzevsky, *Frederick Law Olmsted and the Boston Park System*, Cambridge, MA: Belknap Press, 1982.

8 The detail referred to, as well as the overall sheet, is reproduced in *Works of Lawrence Halprin*, Tokyo: Process Architecture, No. 4, 1978, p. 170. See also p. 180 for sections and construction details of the larger Auditorium Forecourt Fountain, now known as Ira's Fountain, two blocks away.

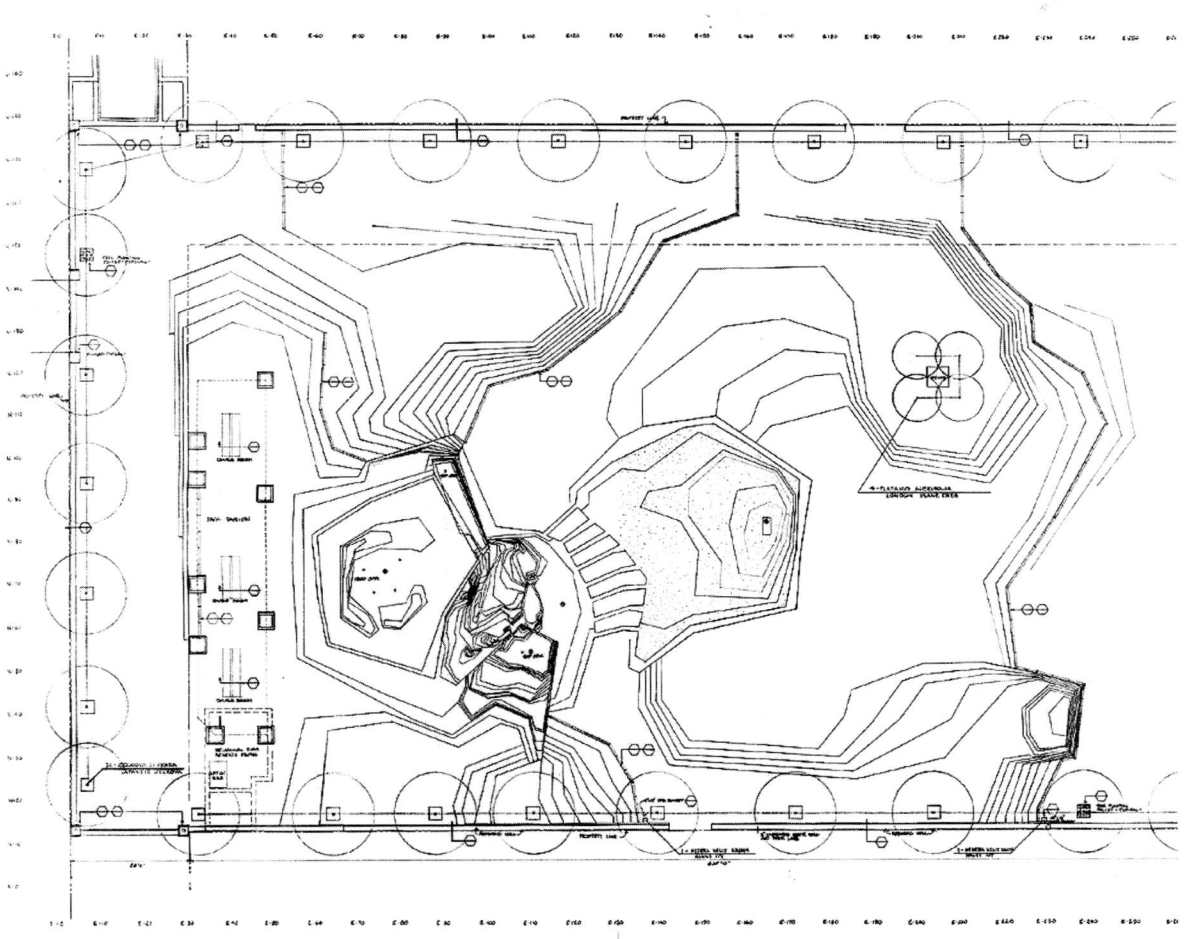

9-11
Lawrence Halprin. Lovejoy
Fountain, Portland, Oregon,
1967. Construction plan with
planting added to layout plan.

*Kroiz Architectural Archive, School
of Design, University of
Pennsylvania*

10

Peter Walker

Modeling the Landscape

For hundreds of years, architects used physical models to study the complex three-dimensionality of their work, as did students of architecture who used models in conjunction with other graphic forms such as sketches, perspectives, and technical drawings. In contrast, over the years, landscape architects relied to a greater degree on mapping or plans of various scales to study and represent their ideas and proposals. It was generally felt that plans were sufficient to represent spatial relationships, grading and layout, and the disposition of plant materials—primarily because the horizontal dimension of landscape is generally considered much more important than the vertical dimension. Maps, plans, and very occasionally a section, remained the standard tools in the past century.

In the late 1960s, The SWA Group started a new program, a summer session for students interested in landscape architecture. Over the next twenty years, students from many universities were chosen for the firm's eight- to twelve-week program, and for the first few years a few arts-oriented high-school students were also included. The internship attracted an amazing array of bright and strongly motivated young people with diverse backgrounds and a wide range of representational skills—that is, from quite sophisticated to virtually non-existent. And it was because of this range that we began to emphasize mechanical representations, namely, grading plans, site plans, sections, and models. This use of models was somewhat odd, for at that time our office rarely used models to any great extent, either for study or exposition. The results in the summer program were amazingly good. In just a few short weeks, everyone, even our high-school students, were thinking and explaining in three dimensions.

In 1975, I began teaching at the Harvard University Graduate School of Design, and many of these summer experiences were the basis of my first studios. From the beginning I required models both as part of the design process and also for use as presentations at reviews.

For almost a century, the Land Grant schools had produced the nation's landscape architects in four- or five-year undergraduate programs. Only a few of these Bachelors of Landscape Architecture (BLAs) continued toward a masters degree, which consisted mostly of advanced design with some additional study of history and theory. At about the same time I went to teach at Harvard, students with no previous landscape architecture degree began to outnumber the BLAs who had been the mainstay of the department's population. These "three-year" or "first-professional-degree" students were generally more broadly educated than the BLAs, with developed skills in research and writing and possessing much wider cultural backgrounds.

10-1
Office of Peter Walker Martha Schwartz, landscape architects; Ricardo Legorreta, architect. IBM Southlake Campus, Solana, Texas, 1984. Model of freeway intersection.

David Walker, Peter Walker and Partners

But suddenly we were faced with the problem of taking quite mature students from many different backgrounds and, in only three years, turning out Masters of Landscape Architecture who expected to practice.

The majority of the "three-year" students had little experience in technical or free-hand drawing. In the first years it became apparent that they could easily learn mapping and mechanical drawing. However, drawing in perspective well enough to represent spatial design ideas was another matter. The department offered special drawing classes, both in the summer before entrance and throughout the curriculum. Yet, without a previous background in art, many "three-year" students could not acquire drawing skills at the necessary level of proficiency. Almost immediately we faced the criticism that these students were not properly prepared to enter professional landscape practice.

In 1978, I became chairman of the department and, partly because of this criticism, I decided to revise the three-year curriculum to the extent of transforming the first year into what has been dubbed (not by me) the "boot-camp" year. In a series of twelve design problems of increasing complexity, each project required mapping, some form of plan, careful sectional representation, and a three-dimensional model. The students developed a clarity that not only allowed them to think almost immediately in three dimensions, but also to move through their lessons with greatly increased rapidity. In just a few months, they were able to express a range of increasingly complex and sophisticated solutions that they could easily present and discuss. Almost everyone with some native artistic ability could make simple models that represented design thinking. By the end of their three years, they were producing work of professional quality.

Since 1976, in addition to teaching, I had been conducting a small experimental office under the umbrella of The SWA Group. Together with five or six part-time students and recent Harvard graduates, we entered competitions and took on small projects with an emphasis on design research. As students came into the office, they brought with them their modeling skills, and these models in turn became the heart of both our investigations and our presentations.

We began to discover new aspects of modeling—what elements could be abstracted, what elements had to be exact and visually clear, what elements needed to be enlarged in order to be useful, and how human scale could be grasped or lost. We learned to make careful, precise model studies of combinations of design vocabulary that could then be shown in larger overall models. We learned that with very careful detailing a 1" = 20'-0" scale model could be looked into at pedestrian level, that is, under the trees.

10-2

Peter Walker and Partners, landscape architects; Helmut Jahn, architect. Sony Center, Berlin, Germany, 1994. Study model of trees, furniture, and paving.

Dixie Carillo, Peter Walker and Partners

10-3

Peter Walker and Partners. University of California, San Francisco, Mission Bay Campus, 2006. Study model of site and landscape, 1999.

Peter Walker and Partners

Sixteenth- and eighth-scale models made this scrutiny even easier. Of course, we also learned that the detail needed for quarter-scale or larger was quite difficult and expensive. To achieve a sense of reality, these larger scales were very time-consuming, and we often solved this problem with models of details rather than models of the entire project area.

We had always known that many of our clients or reviewers could not read plans or deal with abstract representations, but here was a combination tool that could represent three-dimensional space, scale, and usability as well as color, texture, detail, and character. Clients loved models, and we were increasingly able to gain acceptance of quite sophisticated concepts—concepts that were difficult to show in plans and sections or to express verbally or in writing [10-1].

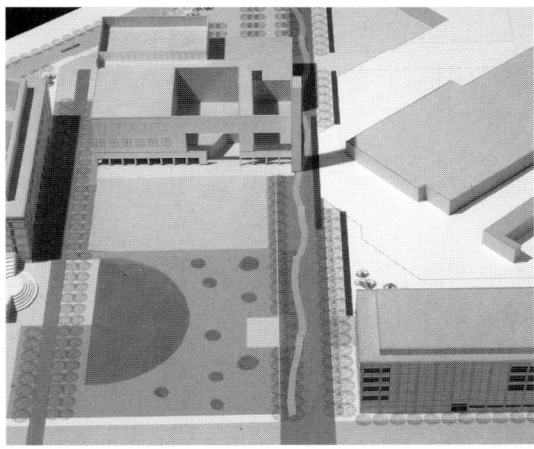

Among the expanded resources of firms in the 1970s and 1980s was a full-time photographer to document new sites, photograph finished work, and assemble slide files of historic projects, vernacular and designed landscapes, and the pieces of landscapes that we call elements, for example, paving, benches, lights, plant materials, and so forth. In the exposition of our designs, we wove these photographed examples into the story of the design, using them to express both existing as well as proposed designs. As our firm grew, much of the work was sited in different parts of the United States and then overseas, both in Europe and Asia. Slides were light and easily put into brief-cases. Our jobs were now running for at least four, and often as long as eight years. We were making many presentations both to private clients and to an increasing number of public review agencies. Slide presentations could be easily modified for political or informational reasons, and they could grow in complexity throughout the life of a project.

Soon we began to make models to be photographed [10-2]. These models treated the overall designs, but also important detail areas. We learned how modeling could be combined with photography to create an understanding of perceptual reality. Sometimes the models in the photographs seemed to be the reality, and we had difficulty telling them from the finished work. We learned to make models that were compatible with plans, sections, professional sketches, and color-slide images of real places and landscape materials —representations that then could be organized into PowerPoint, photographic slide, and board presentations, and printed publications.

Our models often start out at the very early stage of program and scale exploration. They may be little more than existing topography with crude building massing and collaged alternative program representations, what we once illustrated with bubble diagrams and diagrammatic site studies

10-4

Peter Walker and Partners. University of California, San Francisco, Mission Bay Campus, 2006. Presentation model.

Peter Walker and Partners

10-5

Office of Peter Walker Martha Schwartz, landscape architects; Mitchell Giurgola, architects. IBM Westlake Campus, Solana, Texas, 1984. Study model of stream.

David Walker, Peter Walker and Partners

[10-3, 10-4]. The advantage of these models is their ability to create an awareness of scale and spatial reality for the designers [10-5]. Every professional in our office has participated in this activity. It allows each of us to be hands-on in the design process.

For the Nasher Sculpture Garden in Dallas, Texas, for example, PWP produced a series of eighteen models. First came several simple abstract models at thirty-second and twentieth scale that explored in three dimensions our early design proposals [10-6]. These were made of cardboard and construction paper. Then five alternative models studied combined concept alternatives by the architects—Renzo Piano Workshop—and our office [10-7]. These included representations of various tree planting combinations at the scale of 1"=16'-0".

After a general direction was established, we prepared a series of detail models to study individual elements such as pools, stones, benches, lights, and the stepped garden [10-8, 10-9]. These models were built at scales ranging from 1/4"=1'-0" to full scale. Two subsequent presentation models included careful renderings in wood from the Renzo Piano Workshop along with realistic types of tree. These models were used for public announcements of the project. Next followed exact replicas at a scale of 1/16"-1'-0" of the twenty-five to thirty sculptures that were being considered for the opening exhibition. These models assisted in curatorial selection and placement of the sculptures.

10-6
Peter Walker and Partners,
landscape architects;
Renzo Piano, architect.
Nasher Sculpture Garden,
Dallas, Texas, 2003.
Early study model, detail.
Peter Walker and Partners

10-7
Nasher Sculpture Garden,
Dallas, Texas.
Study model.
Peter Walker and Partners

Finally, we made an additional presentation model that incorporated the building housing a work by James Turrell and the major sculpture placements for use in public presentations during the construction process. The last five of the large models were of sufficient detail (and included model people) so that photography could accurately show the scale and quality of the range of garden spaces, including their exact grades and their relationship with the museum building.

Our preoccupation with the pedestrian view provided by models did exclude the visual overview, today demanded by increased public awareness, thanks to airplane travel and aerial photography (see 10-4, 10-7). The overview allows another kind of intimacy and conceptual clarity, one that was once captured in the "above the clouds" views of Japanese paintings and medieval axonometric drawings. Modern high-rise urban development has also put increasing emphasis on viewing the landscape from above. And what about sections? Using developed computer techniques, we have been able to increase their reality and expository quality [10-10].

Using these developed design techniques we can jump in scale, go inside the project, "walk" around. Models, which we began to use with our summer students and then transferred to the Harvard Design School, are but one of several methods that encourage the viewer to enter into that magical world that we try to make here on Earth. And, of course, the first entrant must always be the designer.

10-8
Peter Walker and Partners, landscape architects; Renzo Piano, architect. Nasher Sculpture Garden, Dallas, Texas, 2003. Detailed study model.
Peter Walker and Partners

10-9
Nasher Sculpture Garden, Dallas, Texas. Detailed study model.
Peter Walker and Partners

10-10
Nasher Sculpture Garden,
Dallas, Texas. Section.

Peter Walker and Partners

11

Kirt Rieder

Modeling, Physical and Virtual

In the early 1980s, the State of California commissioned the landscape architects Hargreaves Associates, the artist Douglas Hollis, and the architect Mark Mack, to transform an existing parking lot fronting San Francisco Bay into a landscape for public recreation. During the client interview, the design team of landscape architect, artist, and architect intentionally scrambled their presentation slides to integrate the work of the three distinct entities and to strengthen the perception that the team would be both interactive and collaborative.[1]

Rather than relying on traditional graphic techniques that render collaborative design unwieldy, Douglas Hollis suggested the use of a sizable sandbox to study various developmental strategies. A 5' x 7' sandbox was constructed in the Hargreaves Associates office that supported rapid, if crude, collaborative work that resisted control by any one member of the design team [11-1].[2] The result was an active and immediate relationship between thinking and making that allowed for significant revisions, rapidly proposed and implemented.[3]

Using the sandbox, the time invested in the design studies could be minimized, yet the risk of disruption was high when compared to more stable model-building materials. Keeping the various schemes intact proved difficult as the sandbox occupied a table within a small design office. Despite these logistical problems, there were positive aspects to the medium. Sand as a modeling medium had the distinguishing characteristic of conforming to a natural angle of repose approximating that of an actual earthwork; this kept the sandbox study "honest" in terms of slope and footprint.[4]

The various sand-generated alternatives were dutifully photographed for use in public and client presentations. The State of California project representatives intermittently attended meetings in the Hargreaves office and viewed the sand model in its various stages. The three-dimensional models were intended to help clients understand and embrace the innovative concepts proposed by the designers. To ensure that interested neighborhood constituencies also understood the ideas behind the park's design, a more durable cork model was made, translating the smooth surfaces of sand into the more abstract wedding-cake terraces of contour models.

The translation from sand to cork unintentionally distanced the presentation of the idea from the original concept. Of course, both sand and cork models are abstractions of the landscape. Sand more closely represents a scaled-down miniaturization of continuous landscape surfaces, while cork models increase the degree of abstraction by eliminating surface continuity, replacing

11-1

Hargreaves Associates, Douglas Hollis, Mack Architects. Candlestick Point Cultural Park, San Francisco, California, 1986.

Sandbox used for design studies of landforms that flank the main lawn of the park.

Hargreaves Associates

it with stacked strata. Each layer of cork rendered visible the otherwise invisible contour lines of the terrain. Of course, contours are themselves an abstract representation of grading used to describe three-dimensional topography in two dimensions by assigning intervals of elevation, typically in one-foot increments. This translation creates difficulties for many clients—and even design professionals—because these contours do not exist in the landscape.

The massive size and weight of the sandbox required that all design meetings take place either on site or in the office, ensuring that the collaborators would make all design decisions collectively, based on site revelations or modeling studies.[5] The sandbox encouraged any individual to test ideas and to gather immediate feedback from the broader team, making the medium truly collaborative. Unlike sketching, where the drawing tends to be of small format, or even the 24" x 36" size of conventional drawings, the larger dimensions of the sandbox fostered broad gestures and experimentation.

The construction of the Candlestick Point Recreation Area began in 1986, but then abruptly stopped on the discovery of a rotted wood pier buried within the project site. The problematic decomposing wood was removed in due course, but the landscape grading had to be recalculated and adjusted to correspond to the reduction in available soil for on-site redistribution. The sandbox supported the study of the revised landform alternatives using the reduced amounts of soil available and the dimensional relationship between landforms.[6]

As the sandbox was not portable, the various stages of the design were preserved only in photographs, and the resulting Polaroid prints became the prime two-dimensional records of the continued revisions. An additional drawback was the difficulty of translating the sand forms into drawing, despite the Polaroids. In this particular case, the sandbox designs were not overly complex and so translating the model forms to two-dimensional drawings was not unusually troublesome. The sandbox design studies were inherently low in complexity, exploring proximity relationships and scale comparisons for the various landforms although it did not effectively model the precise intersections of two or more landforms. The days of the sandbox were numbered.

CLAY

As a design and artistic medium, clay has been used for a broad spectrum of purposes from ceramics to modeling concept cars. The sculptor Isamu Noguchi once stated, "Any medium, after all, is new (or old) in time."[7] Noguchi "worked with the malleability of the clay, exploiting the speed with which new shapes can be fashioned and transformed, and freely drawing on the wet surfaces," reveling in the quick results of the medium. He worked with various clays, though primarily for eventual firing for ceramic production.[8]

Noguchi worked with sand too, notably on his *Sculpture to Be Seen* from *Mars* (1947).[9] Many of Noguchi's early landscape models were finalized in white plaster, and some of these were later cast in bronze.[10] However, the smooth surfaces and neutral color of these same Noguchi models were suggestive of working with a malleable material such as plasticine. Clay, unlike plaster, can be rapidly altered and repeatedly re-worked, although water-based clays dry out and crack if left unfired. Roma plastilina is an alternative oil-based clay containing sulfur that remains permanently pliable. Roma plastilina Grey-Green No. 2 is the most widely favored of the four commonly available consistencies, leaning toward the softer end of the hardness spectrum, and referred to generically as clay. While Roma plastilina does remain pliable, it tends to harden slightly over time, thereby maintaining the same heft and weight as the raw form. If not adequately protected, it can accumulate dust that muddies the smooth surfaces and is not always easy to remove. Clay took the place of sand in the landscape office, for use in modeling organic topography, as it had found an earlier place in architecture offices for exploring structural components and details.[11]

Given that practicing designers teach design studios and technology courses at universities, clay periodically appears in various architecture and landscape architecture design courses, particularly at Harvard University's Graduate School of Design.[12] Clay is well suited for quick volumetric studies as well as immersive fundamental skills workshops, as it allows for repeated adjustment and editing by both student and instructor.[13] In the first semester of a landscape curriculum in which students wrestle with a variety of overwhelming concepts and techniques, the use of clay may appear as a return to a more familiar, fundamental skill of making things learned as early as in elementary school.[14]

ISSUES AND METHODS

Clay modeling helps designers explore the two fundamental grading issues: slope and intersection. Slope refers to the geometric vertical rise and horizontal run of a given elevational difference between two points, and is used to describe topography as a percentage, degree, or proportion. Slope can also describe a section through a landscape, and is particularly easy to record in a conventional orthographic view. Applying a consistent slope throughout a three-dimensional landscape becomes more challenging as a designer juggles scale and proximity of one or more volumes to each other, in addition to maintaining the parameters that define pure geometric volumes. An added level of complexity arises when depicting the overlap of two or more solid volumes, resulting in a Boolean union with a predictable intersection of their respective surfaces.[15] For instance, conventional projection drawings can capture the intersection of a cone and a pyramid with a square base and four triangular sides—but the drawing does not yield a three-dimensional model nor lend itself to rapid adjustment. The intersection between these two volumes becomes infinitely variable as the two solids are rotated, scaled up or down, or positioned in different locations, leading the designer to seek a medium for testing multiple scenarios in a minimum period of time and with the least effort. The resultant volume can be understood only from the parameters that defined the two original solids: dimension, consistent slope, and uniformity of surfaces, in addition to the unique points of intersections.

The desired end result of designing a landscape is not a set of drawings, however, but a built landscape. Modeling the landscape in three dimensions identifies areas needing further adjustment. The apparent diversion from clay into geometric modeling leans heavily on designer-determined guidelines or rules for executing the work, recognizing that the three-dimensional model will be translated back into the drawings provided by the contractor for construction.

WORKSHOP

Harvard's Graduate School of Design Clay Landform Workshop set three objectives: (1) the development of a basic landform vocabulary; (2) the use of landforms to define spaces; and (3) the acquisition of quick modeling skills to aid in the preparation of grading plans. The first objective helps students develop a vocabulary of fundamental landforms including constructed geometric volumes and those derived from natural processes. This vocabulary ranges from primitive cones and pyramids to glacial drumlins and Aeolian sand dunes.[16] The elements of this vocabulary do not suggest that a cone, drumlin, or any other landform is preferable for student adoption, however, but rather that the range of diversified forms will expand the student's ability to explore and describe simple volumes [11-2].

The studio's second objective was to "exercise and build design muscles," developing a familiarity with, and proficiency in, crafting three-dimensional volumes as the primary form and space generators of a designed landscape.[17] This emphasis on the use of landforms to define space is distinctly landscape architecture-focused, as is the parallel use of vegetation.

The third objective develops the student's ability to confidently conceive and rapidly configure landforms from which a two-dimensional grading plan evolves, as well as the opposite path: the visualization of three-dimensional surfaces and volumes from a grading plan. The overarching goal is to conceive a designed landscape that can be tested in model form, documented in drawings, and ultimately constructed. Notably, students entered the workshop prior to taking classes in which they would learn the technical conventions of contour grading and site sections.

The intent was to immerse students in the process of creating earthworks before considering the practical grading skills and conventions they did not yet understand. As such, the workshop offered valuable trial-and-error exposure to grading concepts, preparing students for the less forgiving technical aspects of shaping the land, and differentiating existing topography from designer-initiated grading.[18] The fundamental message defined landforms as intentional and calculated constructions, configured according to "rules" determined by the designer rather than only a surprising coincidence of the tidy resolution of contours and ridges.

11-2

Landscape student, Harvard University Graduate School of Design Clay Landform Workshop, 2001.

Geometric and geologic volumes, both additive and subtractive, layered together yielding a transformative landscape.

Kirt Rieder

The workshop set a number of ground rules, including the stipulation that no slopes might exceed 2:1, although this proportion of horizontal to vertical exceeds the natural angle of repose for most materials. Other rules required that each subsequent shape intersect with one or more of the preceding shapes, that all intersections be sharp and uniform, and later, that subtractive operations resulted in the removal of clay from previously configured shapes.

The ability to describe individual land components allowed the larger composition to be understood as either a collection of recognizable additive and subtractive shapes, or as a complex composition of surfaces and intersections. This transformative topographic composition was no longer subject to facile verbal description, but still abided by a series of basic rules relative to slope and continuity of lines. Blurring intersections or blending the shapes into the base was discouraged for this exercise, because the identity of the shape rapidly diminishes when boundaries and intersections are intentionally smoothed out. The clarity of the study depended to a large degree on the student's skill at modeling the medium. If the initial 12" x 12" base of clay was rough, and the subsequent shapes poorly formed, it became more difficult to model abstractly and to precisely resolve the contours of the surfaces. The emphasis on distinct forms and pronounced intersections between these surfaces runs counter to the prevailing attitude in landscape architecture to "soften" or blend grading into the existing conditions to make new interventions appear seamless or solely as background scenery. The workshop emphasized "hard-edged" landforms to force students to wrestle with the

challenges offered by geometric and intersecting surfaces as a means of developing the rigor to approach and solve later terrain problems.

In reality, sharp ridges and smooth surfaces are necessary to accentuate grading for landforms to be legible within the context of adjacent structures and densely vegetated surroundings. This attitude by no means reflects only a twentieth-century sensibility. The Hopewell Indians in Central Ohio constructed earthworks and effigy mounds as precise geometric shapes that remain powerfully distinct from the broader landscape to this day [11-3]. The crispness of these earth forms caught the eye of early settlers from the 1790s through the 1840s, encouraging their documentation prior to subsequent destruction by later agricultural development.[19]

11-3
Mound City National Monument, Chillicothe, Ohio, 1998.

Hopewell cultural complex, dating from between 200 BCE to CE 500, comprised of more than a dozen conical landforms within a square berm.

Kirt Rieder

The end goal of the workshop was not so much a good-looking model, but a constructed, scale-less landscape composition with dramatic contrasts between the existing ground plane and the student's intervention. Over nearly a decade of the workshops, the students demonstrated that their mastery of solving complex geometries in abstract clay studies afforded greater understanding and facility in solving basic grading and drainage problems in subsequent design studio and technology coursework.

MEDIUM

Clay became the preferred physical modeling medium because it is responsive, plastic, forgiving, and easy to work with. The clay model can be tilted and rotated for an infinite number of viewpoints. On the other hand, clay demands a fair degree of attention to craft, especially when creating straight lines and smooth surfaces. Another downside of clay is a sticky residue that clings to the hands and equipment and its lingering smell.

Three-dimensional clay models are typically more accessible to a broader audience than three-dimensional computer models because their physicality makes them easier to understand through touch and sight. In contrast, the projected two-dimensional digital image of a landscape still leaves many viewers struggling to form a mental image of the design concept in the absence of richly detailed textures that provide crucial cues to depth and distance. Clay supports free inspection whereas digital models require con-

trolled vantage points. And while three-dimensional computer images offer infinitely variable station points and perspectives, revisions to digital models often take more time to execute than clay models. While clay has the immediacy and tactility lacking in digital models, the latter are more easily converted to various conventional drawings with multiple end uses, including those of presentation and construction. However, the development of fundamental clay-modeling skills accelerates the adoption of more abstract and structured mathematical concepts necessary for executing digital computer modeling.

TRANSLATION

Craft

The translation of a scale clay model to a constructed landscape relies on craft. Craft in this context refers to planar surfaces that appear neither bulging nor sunken, and uniform lines that neither sag nor jog erratically. Craft fosters the discipline to precisely and persuasively describe a proposed landform. As one example, Hargreaves Associates initially worked in clay to develop a basic concept for the annual Festival of Gardens at Chaumont-sur-Loire, France, snapping Polaroids to record ongoing design evolutions over a period of days [11-4]. A final clay concept was packaged, and shipped to France for approval. The presence of the clay model persuaded the competition administrator, in the designer's absence, of the proposal's clarity though it was unusually complex for a staff unfamiliar with constructing earthworks. Once approved, construction documentation commenced.

11-4

Hargreaves Associates. *La Terre en Marche*, Chaumont-sur-Loire, France, 1995. Final clay study for client review.

Hargreaves Associates

11-5

La Terre en Marche, Chaumont-sur-Loire, France. Completed installation.

© *Yann Arthus-Bertrand / Altitude*

For this phase of the project, a model less than a square foot in area was the sole design tool. Translating the three-dimensional clay model into a set of two-dimensional construction drawings was a crucial step toward realization. Drawing on the principles described in the clay landform workshop, each shape was solved individually during the process of documenting the proposed terrain.

Using computer-aided design (CAD) software, each shape was described by defining ridges and contours prior to stitching the components together. Contour spacing for each volume was determined by setting a desired slope independent of existing adjacent topography and by offsetting, at an equal distance, all subsequent contours. The grading process integrated the contour lines of intersecting volumes resulting in a complex landform described by contours and the junctions of the ridges and valleys. This process resulted in a composition of three primitive landforms fused together, yet decipherable as independent volumes nonetheless [11-5]. A grading plan is not the ultimate goal. Models and construction drawings were the stepping-stones to a built landscape.

Interpretation
Transforming clay to drawings involves considerable interpretation due to the differing degrees of precision afforded by hand-crafted clay models and CAD-constructed drawings [11-6]. Considerable adjustment is necessary to complete the process for rendering in drawings what appears so clearly in

a clay model. In the 1990s, students photocopied small clay models to use as crude two-dimensional bases for their grading plans. Not surprisingly, this mode of representation was both messy and contentious for the institutions and for other students who used the same machine for other rightfully intended uses. More recently, digital photos of larger clay models have made the translation process much quicker, more precise, and significantly less messy.

A raster image can be inserted into a digital survey file to scale and adjust the forms to correspond with the prior vector data. The grading process focuses on assigning contours and uniform slopes to each volume defined in the raster image. Individual landforms are graded from the apex downward, then adjusted and integrated with the existing contours, and where multiple volumes intersect, adjustment is normally necessary. The resulting grading plan effectively integrates the once abstracted volumes into a scaled drawing that can be used for calculating projected soil volumes and surface areas for planting, irrigation, and drainage.

Scanning

More precise, though costly, translation tools are currently available. Three-dimensional laser digitizers can now record a "point cloud" of an object or model, capturing its outline and surfaces with literally millions of data points. These points are integrated into a three-dimensional digital file more accurately than either by point-by-point hand digitizing or by photographing and assigning contours after the fact. Today laser digitizers are beyond the financial means of most firms, however, although some schools have already purchased them and their use is increasing. Clay landform models are scaled representations of a larger landscape and, so, even in the case of a finely crafted model, there is an inherent lack of precision in translating a scale model up to a 1:1 constructed landscape. The interim step of drawing and adjusting contour lines remains a necessary step despite the overwhelming data-recording devices.

In the end, if clay models are used to explore a concept, why not bypass clay altogether and generate a three-dimensional digital model? Working with clay provides a more fundamental and exponentially faster entry point to studying volumetric landforms, free from the governing conventions of grading plans, and is far less daunting than a blank computer screen. Experience at Hargreaves Associates has shown that a three-dimensional digital model is more successful and useful for representing a design derived from a two-dimensional CAD file—and not the reverse.

11-6

Hargreaves Associates.
Expo '98, Lisbon, Portugal,
1994.
Clay competition model.
Hargreaves Associates

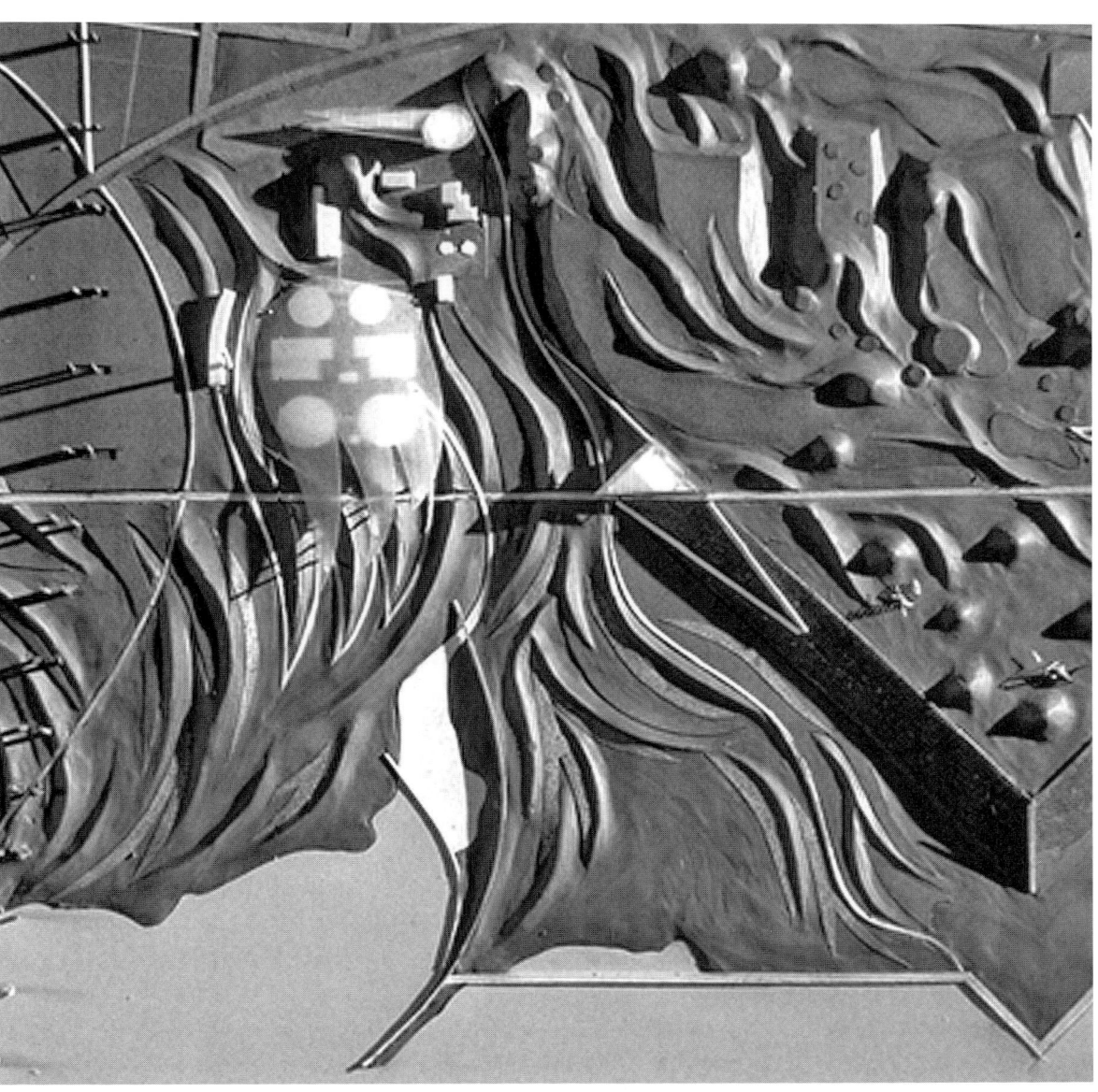

Simplicity

Determining a consistent distance between contour lines in a two-dimensional construction drawing sets uniform slopes that can be staked and built in the field, whereas working with three-dimensional digital drawings at the outset requires the additional step of revising contour intervals to whole-integer dimensions [11-7]. Good construction drawings are simple drawings, readily understandable to the various trades that construct the landscape. Therefore, there is a preference for working with whole numbers rather than odd fractions, with reasonable tolerances of not less than one-hundredth of a foot.[20]

Allocating time and money to exploring a landscape in clay during the earliest stages of a project may seem to be a luxury. However, Hargreaves Associates has found that the resources are well spent on a three-dimensional kick-off to the design process, concurrently integrating the programmatic and circulation patterns into the studies of the volumetric landscape. Although this entails adjustment to the modeled landscape, the project is often better integrated by working simultaneously on all aspects of the design rather than solving the grading at a later time. This early focus on topography illuminates the relationships between program and circulation that might otherwise be overlooked, had the evaluation of topography been delayed.

In their work, Hargreaves Associates strives for a legibility of landforms by accentuating the contrast between the pre-existing ground plane and the reformed configuration [11-8]. This is achieved by crafting uniform surfaces

11-7

Hargreaves Associates. Guadalupe River Park, San Jose, California, 1995. Final staking and grading operations for floodplain landforms.

Hargreaves Associates

11-8

Hargreaves Associates. Crissy Field, San Francisco, California, 2001. Landforms contrasting with "flat" ground plane.

David Sanger

and slopes and sharpening the intersections between planes and volumes.
Bypassing sketching, or clay, in favor of study solely on the computer delays
the recording and testing of the initial concept until late in the design
process which may reveal unwanted conflicts between programmatic and
topographic relationships.

PARALLELS

In spite of the promise of an all-digital design process made possible by
falling hardware prices and software tailored for describing complex objects,
clay remains a valuable medium for design exploration. Examples from seemingly unrelated design fields demonstrate that the use of clay remains a
significant design tool. Recent technology now facilitates the translation of
clay to CAD, strengthening the prominence of the initial medium by expediting
the transference of clay to the design drawings and subsequent fabrication.
Two examples are taken from automobile and basketball shoe design. BMW
continues to build clay models of their cars at full scale when developing new
prototypes. These clay models are then laser digitized and further developed
using the computer. Precise milling equipment, accurate to within one mil-
limeter, executes any revisions and adjustments to the clay model. Thereafter,
the clay models are painted to convincingly appear as finished products,
although the design process actually continues. BMW prides itself on hand-
craftsmanship and integrates the clay models into the design process not
only in the earliest stages, but also for feedback when testing the physical
characteristics of the design. BMW strives for *Flachengenauigkeit*, translated
as the "precision of the surface," best achieved through sculpting the surface
by hand.[21]

11-9
Morphosis and Hargreaves
Associates. NYC2012 Olympic
Village, New York City, 2005.
Collaged digital image
integrating architecture
and landscape into a site
photograph.
Hargreaves Associates

At the Volkswagen and Audi Design Center in Simi Valley, California, nearly
every new model is developed in clay because skilled modelers "still produce
a more visually attractive surface with their hands."[22] A realistic model is
of paramount importance as many reviewers participating in design and
product evaluation are neither designers nor involved in the day-to-day
development process: the model has the responsibility of convincing these
decision-makers, using surfaces that appear real. The Design Center also
uses laser scanning, translating a cloud of millions of points into a CAD file
for further development. In one of the more unusual examples of clay use,

the 2000 design for the Adidas Kobe Bryant basketball shoe was designed by Audi, which modeled the shape and surfaces of the shoe using clay models. Coincidentally, the Audi TT automobile provided a literal, if odd, inspiration for the actual shoe design. Car and shoe were crafted in clay to express directional, forward movement.[23]

Both examples support the argument that despite recent technological advancements in imaging—even with the seemingly unlimited financial resources of leading automotive and shoe corporations—clay retains its role as a critical step for recording initial concepts. Notably, General Motors' market share and revenue have continued to erode even as they trumpet their discarding of clay models and adoption of digital-only design of new models.[24] The ability to produce, and resources available to model, cars, shoes, and landscapes entirely in a digital context already exist and are financially feasible, yet there is a basic sense of control, craft, and satisfaction that results from a physically, rather than digitally, made model.

3D Landscape
Hargreaves Associates uses Auto-des-sys Form*Z and Autodesk 3ds Viz to model many types of landscapes, particularly for perspective views of proposed designs inserted into photographs of the existing context photographs. These also test the spatial relationships within the design concept and the efficacy of the design [11-9]. Individual shapes or components are modeled independently and knitted together to form a complex landscape.

Despite the initial time investment, an infinite variety of views—perspective or axonometric—can be generated for internal design team study and client review after the basic information has been entered. However, at times, clients have been confused or put off—rather than persuaded—by the abstract shapes and colors of computer-generated images. The well-known complaint is that the three-dimensional views are too cold, too stiff, "too Edward Scissorhands," referring to the often over-saturated color output images.[25] As with car design, there is always the risk of unintentionally damaging a fledgling concept by presenting a computer rendering to a client too early in the process, without adequate graphic textures and accurate shadows, or by failing to convey the objective of the studies to less design-aware clients.

In an effort to make the digital models more accessible to a lay public, considerable effort has been channeled into Adobe Photoshop effects and texture-mapping to heighten the realism. "Texture-mapping" is the digital process of applying a pre-selected image as a continuous pattern onto a three-dimensional digital surface or two-dimensional Photoshop image, suggesting changes in surface and light. Texture-mapping helps clarify the illustration by assigning textures and atmospheric character to surfaces such as grass or stone. As significant time and effort are expended on digital tasks such as these, resistance grows to changing or eliminating an option. Designers become invested and committed to a particular concept as they pour increasingly long hours and effort into detailing an environment. The relationship between time and resistance to change occurs in both physical

and virtual models, but can be more problematic when texture mapping and manipulation of color, scale, and pattern obscure an underlying structural problem with the basic concept.

SUMMARY

There are several ways to study concepts for developing landforms including traditional section drawings and grading plans. Studies in various media provide feedback for testing, revising, and generating ideas: in paper, in chipboard, in clay, digitally in two dimensions, and digitally in three dimensions. To efficiently focus energy, Hargreaves Associates has used clay models as an economical and quick study material easily evaluated and altered in real time. The computer solves the precise geometric intersections between volumes and sets uniform contours to accurately represent a consistent slope. Contours describe spatial relationships between landform volumes to each other and to the adjacent existing landscape. CAD is more heavily utilized for subsequent aspects of design development and continues through construction drawings. Form*Z and 3ds Viz facilitate the study of perspective views and are shared with clients and public for endorsement of an evolving design.

Despite the primitive simplicity of clay and the increasing sophistication of digital image files, Hargreaves Associates often commissions hand-drawn renderings for public presentation or donor fundraising, such as those by the San Francisco architectural illustrator Christopher Grubbs [11-10]. These renderings may be derived from a basic wire frame, three-dimensional computer image provided by the firm, or from perspectives constructed by the illustrator's hand. Colored pencil and watercolor renderings are less abstract and alien than the cool rationality of the computer-generated image and are perceived as more "real" or warm, more filled with life. Notably, Grubbs prefers to construct his perspectives by hand rather than from a computer-generated wire-frame drawing, as this allows him to establish the cone of vision and choice of vantage point to compose the best perspective, and ultimately to show the proposal to its best advantage.

In the end, clay models best describe the landscape volumetrically and in relation to the existing landscape. Clay remains the fastest, most dynamic, and

11-10
Hargreaves Associates. Grant Park Master Plan, Chicago, Illinois, 2003. Pencil and watercolor rendering depicting landforms and how they are used by pedestrians.

Christopher Grubbs and Hargreaves Associates

most forgiving modeling medium. Despite the rapid progress of increasingly sophisticated terrain modeling software, clay retains the appeal of physical construction through direct contact. Clay models allow the designer and the viewer to see all of the site at once, from any vantage point, while simultaneously focusing on specific areas of interest, in a way that digital models and renderings have yet to achieve. Clay models also serve as a springboard to CAD drawings and digital three-dimensional models, as well as to hand-drawn perspectives, and remain the most versatile tool for testing landscape grading strategies.

EPILOGUE

In 2002, Tangible Media investigators at the Massachusetts Institute of Technology's Media Lab invited the author and select Graduate School of Design faculty and students to attend two preliminary demonstrations of "Illuminating Clay." The investigators had developed an interface that allows users to alter the topography of a clay landscape model, capture the changes with a laser scanner, and project the results back onto the actual clay surface as a series of analytical studies—all in real time. One objective was to develop an intuitive interface linking the vast quantities of computational data gleaned from geographical information resources and the tangible immediacy of changes to a physical model.[26]

By late 2004, "Illuminating Clay" had spawned a subsequent project "Sandscape," whose continuous Tangible User Interface (TUI) allowed users to scan a sand landscape model, view resultant computational analysis, and project it onto the model's terrain in real time. The switch from clay to sand was triggered in part by the switch from an expensive laser scanner to a less expensive infrared camera that could read the depth of the silica material. The project noted the increasing complexity of new technologies and the increasing accessibility to a wider range of disciplines, yet acknowledged that a tangible user interface "takes advantage of our natural ability to understand and manipulate physical forms while still harnessing the power of computational simulation." The simulations allowed active topographic manipulation paired with immediate, resultant graphic landscape analysis including slope, elevation, orientation, shadow, and water flow. One additional result of "Sandscape" was that TUIs "seem valuable platforms to communicate design decisions to non-experts, allowing them to become involved in the design process" which is integral to reaching a consensus for constructing a landscape.[27]

NOTES

1 Conversation with Glenn Allen, Principal, Hargreaves Associates, Cambridge, MA, April 2000.

2 Conversation with Douglas Hollis, Berkeley, CA, May 2000.

3 Conversation with George Hargreaves, Cambridge, MA, April 2000.

4 Conversation with Glenn Allen, Cambridge, MA, April 2000.

5 Ibid.

6 Ibid.

7 Isamu Noguchi on his life and work, in Isamu Noguchi, *A Sculptor's World*, New York: Harper Row, 1968. Excerpted, online. Available at: http://www.noguchi.org/intextall.html#sculp.

8 Bruce Altshuler, "The Ceramic Sculpture of Isamu Noguchi," in *Isamu Noguchi and Kitaoji Rosanjin*, Tokyo: Sezon Museum of Art, 1996. Available at: http://www.noguchi.org/baceramic. html.

9 Kenneth Frampton, "Sculpture in a Commemorative Landscape: Louis Kahn and Isamu Noguchi," in *Play Mountain: Isamu Noguchi + Louis Kahn*, Tokyo: Watari-um, 1996. Available at: http://www.noguchi.org/frampton. html.

10 Ibid.

11 In the mid-1980s through the early 1990s there were few computers in landscape architecture offices, and widespread adoption of computer-aided drafting (CAD) software had yet to occur. The use of clay, Polaroids, copy machines, and tracing preceded the use of CAD. Architecture offices such as Antoine Predock, to name just one, used clay through the 1990s although it had surfaced elsewhere earlier.

12 In 1992, George Hargreaves—as a visiting critic—initiated an annual two-week-long landform workshop within the basic First Semester Core: Landscape Architecture Design Studio, taught by Elizabeth Meyer, Gary Hilderbrand, and Laura Solano, at the Harvard University Graduate School of Design while the author was a student. This workshop loosely followed earlier model-making workshops led by Peter Walker at Harvard and at the SWA office.

13 From 1996 to 2004, the author taught the workshop, compiling and synthesizing a variety of sources, including diagrams drawn from Douglas Way's *Terrain Analysis*, Upper Arlington, OH: DWA, 1978, a collection of slides derived from David A. Rahm's *Geology Study Guide*, New York: McGraw-Hill Book Company, 1971, and John Beardsley's *Earthworks and Beyond*, New York: Cross River Press, 1989, to name but a few.

14 The landform workshop was viewed by some students as a welcome relief from the more taxing demands of studio due dates, while other students' frustration grew at their difficulty in completing assignments the same day, or overnight if they had fallen behind.

15 "Boolean union" refers to the mathematical operations used in solid modeling and constructive solid geometry (CSG) to create complex surfaces or objects from simple shapes. The use of Boolean unions, intersections, and differences in current three-dimensional computer graphics programs can provide exacting and complex objects, although this description was relatively obscure in the early 1990s. The Boolean union effectively welds or fuses one or more solids together, resulting in a single solid with no deformation.

16 The workshop encompassed man-made shapes defined largely by purely geometric parameters. "Natural" shapes referred to those landforms resulting from glacial, volcanic, Aeolian, and water-influenced processes of erosion and deposition.

17 The syllabus for the initial clay landform workshop, Spring 1992.

18 Conversations with Michael Blier, Harvard University Graduate School of Design core studio instructor, Cambridge, MA, 2000–2005.

19 E.G. Squier and E.H. Davis, *Ancient Monuments of the Mississippi Valley: Complete 48 Plate Collection*, 1848. Image clarified and copyrighted, A.W. McGraw, 2000.

20 Conversation with Catherine Miller, Principal, Hargreaves Associates, Cambridge, MA, 2005.

21 Stefan Thomke, "Managing Digital Design at BMW," Design Management Journal. Spring 2001. Available at: http://www.findarticles.com/p/articles/mi_qa4001/is_200104/ai_n8937247#

22 NVision, "Case Study: VW Scans Clay Models with Lasers," *Quality Magazine*, March 2004. Available at: http://www.qualitymag.com/CDA/Archives/af36995196c38010VgnVCM100000f932a8c0.

23 Audi AG, "The Kobe: Adidas & Audi Collaborate," *AudiWorld*, November 2000. Available at: http://www.audiworld.com/news/00/kobe/content.shtml.

24 Sean Gallagher, "GM Plans Digital Turnaround," *Baseline*, November 2002.

25 Tim Burton's (1990) film *Edward Scissorhands* was set in a highly stylized 1950s subdivision dominated by pale pastel-colored houses.

26 Ben Piper, Carlo Ratti, and Hiroshi Ishii, "Illuminating Clay: A Tangible Interface with Potential GRASS Applications," Proceedings of the Open Source GIS–GRASS Users Conference 2002, Trento, Italy, 11–13 September 2002, p. 2.

27 Hiroshi Ishii, Carlo Ratti, Ben Piper, Yao Wang, Assaf Biderman, and E. Ben-Joseph, "Bringing Clay and Sand into Digital Design—Continuous Tangible User Interfaces," *BT Technology Journal*, October 2004: 295.

12

Marc Treib

Photographic Landscapes:
Time Stilled, Place Transposed

I. INTRODUCTION

The celebrated image of Luis Barragán's stable courtyard at San Cristobal crystallizes the panoply of issues that shape the relation of photography and landscape [12-1]. The photograph represents the vision and the work of a single person: Armando Salas Portugal. In an instant removed from time, this compelling image forever fixes, with light and chemicals, human and animal, architecture and landscape. Like a fly embedded in amber, these horses are set against the brilliant magenta walls for the ages; they will never move again. Nor will the man who leads them, nor the light and shadow that Salas Portugal so deftly captured. Over Barragán's own framing of landscape space, the photographer's rectangular frame recomposes the work of the designer and the life within the setting that has been so beautifully created. The image is an idealized, if not absolutely ideal, moment. With it comes an almost alchemical change: space has been converted into time and converted once again into a two-dimensional representation. For most people attending the exhibitions and reading the books about this architect's work —for those who will never travel to Mexico to experience the sublimity of Barragán's work in the flesh—this photograph will become synonymous with the stables at San Cristobal. More particularly, this one particular photographer, Armando Salas Portugal, has created the Barragán most of the world will know. Is this a problem?

In many ways, the photographic image has reigned as the supreme medium for most of the twentieth century. The image stands in place of its subject, multiplied in the hundreds, thousands, or even millions, and today spread instantaneously around the world through electronic means and digitalization. Of all the means of image representation prior to the computer, photography has been the most potent. In the age of electronic reproduction, the photographic image (and its sister the cinematic image) dominate and coerce how we perceive and live.

While a democratic vehicle of the first order, allowing an almost universal access to information, the omnipresent electronic image has further privileged the sense of sight. We are still told that seeing is believing, although today we well know the extent to which images may be manipulated and made to deceive. This is not to say that the word does not remain a potent force within our culture, today given even broader distribution by the computer and world-wide electronic mail, but its supremacy has been undermined and its foundation more than shaken. And to at least one generation that has matured watching television and the computer screen instead of reading books, the influence of the word has been dramatically muted. In addition, recent efforts at fashioning a "virtual reality" have bypassed the word and

12-1
Luis Barragán. San Cristobal, Los Clubes, Mexico City, Mexico, 1968.

The perfection of the photograph fixes architecture in time; our reading and understanding of Barragán's work are shaped by the photograph.

Armando Salas Portugal, courtesy of The Barragan Foundation, Birsfelden, Switzerland / ProLitteris, Zurich

connected directly with the eye and the ear, consequently reducing the role of cognition in perception.

The photograph was the first medium to "objectively" record our world, supposedly free of human intervention. But, as Susan Sontag has asserted: "This very passivity—and ubiquity—of the photographic record is photography's 'message,' its aggression."[1] Yet it was no accident that for decades Eastman Kodak's guides to popular photography were entitled *How to Make Good Pictures*: we do not passively take photographs; we construct them. Through our choice of time, camera, lens, film, stance, exposure, and chemical develop-ment we manipulate the pictorial fields before us, including those of the designed landscape. Sontag again:

> A photograph is not just the result of an encounter between an event
> and the photographer; picture-taking is an event in itself, and one with
> ever more peremptory rights—to interfere with, to invade, or to ignore
> whatever is going on.[2]

One somewhat-dated history of pictorial representation stressed the evolution of objective—that is to say, conventionalized—representational norms as central to the transmission of knowledge (and hence human development). William Ivins's *Prints and Visual Communication* (1953) chronicled the incremental "objectification" of drawing systems since the Renaissance.[3] Until systems of representation were standardized—according to Ivins—variations in depictions could be assumed, with equal merit, to stem as much from differences in artistic languages and abilities as from differences in the form of the rendered subjects. That is, disparities in a series of drawings could result as much from the artist's perception and technique as from his or her manner of communication. Two drawings of a leaf (Ivins cites the example of the herbarium), for example, each believed to represent the same plant, could appear quite different in their rendering. Views of a city or a garden, each taken by a different artist, could produce images wildly dissimilar. One might follow closely the rules of perspective; another might take liberties with scale and detail; a third might hark back to the pre-perspective renderings of symbolic rather than metric space. The reader could assume nothing and, thus, there could be no true exchange of knowledge or ideas. Subjectivity outweighed convention.

To Ivins, photography represented the culmination of objective graphic communication, untroubled by the subjectivity of the draftsman and the vagaries of the medium. In addition to disseminating pictorial information to a broader audience, the photographic medium was dispassionate and neutral, as an industrial process equally adept at portraying a tree or a natural disaster. Recounted through the cold eye of the camera, photography showed things as they were, not as the individual might perceive them. Or so some believed.

Later writers such as the art historian John Berger reminded us that photography's ultimate objectivity, as seen by Ivins, was a chimera. In *Ways of Seeing* (1972) and in numerous essays thereafter, Berger demonstrated that all images, including photographs, reflect the bias of both the culture and the photographer.[4] While film does record patterns of light, those patterns are affected by the choice of viewpoint and focal length of the lens, the type of film, atmospheric conditions, stability or movement of the camera, and a host of other factors, including those introduced in the darkroom. No, the photographic process might be more industrial or mechanical than the drawing, but it too reflects the subjectivity of the photographer. Although seeming at first to be only a pictorial mechanic, the photographer is, in fact, both an author and an artist.

Jonathan Crary has traced the change in the relationship during the first decades of the nineteenth century between the viewer (recast as a more active observer) and the viewed. In *Techniques of the Observer* (1993), he described a juncture between the belief that the view exists as a complete entity external to the viewer, and one that regards perception as an assemblage of information that must be reconfigured and interpreted by the observer in order to render it coherent.[5] No view is complete beyond the eye; each is instead a mental and personal construct. Michael Podro, in turn, tells how our comprehension of depictions lies between our regard for the surface as painted and the picture as imagined: in his words, "[t]he convergence of …motifs mobilizes the structure of depiction as a metaphor of transcending literal vision."[6] While less evident in the mechanical and smooth surface of the photograph, our ability to read the image engages both these dimensions of depiction.

Thus, we see that the photograph is neither an untouched record of the world, nor is it passive. At its very root, the photograph (normally) extracts the subject from its world and encases it within a (new) rectangular frame. Framing raises the issue of composition because it reconfigures the relationship between the elements that comprise the view, and the part included to that which is left out. In this sense, the photograph bears only a passing relation to the actual ingredients of the world beyond the camera. Within the rectangular slice of the photograph, those fragments that constitute our "reality" have been substantially reconfigured to compose a new pictorial reality.

This introduction sketches only in rather broad strokes several of the issues concerning photography and our world; all of them apply to our depiction of designed landscapes. The space given to this chapter hardly admits treating such a wide-ranging subject in great breadth or its even greater depth. Therefore, of the full range of considerations affecting photography and landscape architecture, only three aspects will be discussed, but three critical to the question at hand. These are reconstruction, time, and displacement.

II. RECONSTRUCTION

For many designed landscapes executed in natural settings, the aesthetic power derives from the articulation of existing conditions, what I once termed an "inflected landscape."[7] The design constitutes an essentially new configuration, a reconstruction that retains certain aspects of the natural order while simultaneously asserting its identity as a construct. The power of even a "soft" geometry as a contrast to the seeming irregularity of the existing order would seem obvious. When these works are photographed, however, an additional displacement takes place: each work is now framed by a rectangle [12-2]. The act of framing in itself can make an artwork from disorder by imposing identifiable limits against which and within which the elements now interrelate. The resulting order is new; it may create, heighten, or diminish the power of the design as intended by its maker. And by establishing the conditions for this recomposition, the photographer functions as an artist regardless of the original subject.[8]

Landscape design may itself embody a form of framing. A garden may be taken as a zone of modified ecological process, at times created with aesthetic intentions.[9] A courtyard garden—whether set in pastoral or urban conditions—derives its primary identity from the order imposed by its

12-2

Seawall, Awajishima, Japan.

Photographic framing creates new relationships that need not exist in the greater world. In particular, the rectangular frame organizes the freer aspects of nature, or allows us to contrast the natural with the designed— or capture framing inherent in the scene.

Marc Treib

12-3

Patio de los Naranjos, Cordova, Spain, fifteenth century+.

The perimeter walls and those of the church provide the primary definition for the courtyard, like the photographic frame of the camera.

Marc Treib

12-4

Ryoan-ji, Kyoto, Japan, c. 1500.

The encircling wall of earth and clay frames the garden of gravel, moss, and stone—set as a void against the vegetation of the surrounding landscape.

Marc Treib

physical enclosure. For example, the encircling wall, and resulting space, of the Patio of the Oranges in Cordova establish a pacified zone detached from the pattern of surrounding constructions which have developed by accretion [12-3]. Like the earthen cuts of Michael Heizer's *Double Negative*, the frame first establishes the locus, the place.[10] Within the walls, the placement of the trees on a grid reflects the dimensions and order of the columns of the adjacent mosque and the garden's enclosing walls. The shared order establishes a continuity between garden and frame, both detached from the "reality" of the city. In Kyoto, on the other hand, the circumferential wall at Ryoan-ji distinguishes the austere contemplative garden from the more naturalistic landscape beyond [12-4]. The seemingly random placement of the stones and moss appears to rebel against the bounded rectangular field that contains them. In both cases, the framing is the first act toward constituting the garden.

With its superimposition, the photographic frame instigates a new order, and as such, creates its own landscape. Consider the relationship between photographic vision and designer's intention. Photographs of Versailles invariably feature the central axis, aligning the Latona and Apollo fountains and the Grand Canal [12-5]. The far and near determine a straight line, almost as one would align a gun sight. Images taken along the axis, playing statues against trees, architectonic order against natural vegetation, all further the commonly held belief that the garden was essentially an axis that extended from the apartments of Louis XIV to the horizon, and then to infinity. Yet that "reality" is but one of the possible readings of the vast gardens of Versailles.

In the literature more oriented to the tourist, the micro rather than the macro scale prevails. To achieve the feeling of the garden as a garden, publications weight photos taken obliquely or closer in, images featuring single buildings, fall color, details in the parterres, or the expressions on the faces of statuary. But how many of these images proffered either for scholarly or tourist consumption avoid the clichés and address the alternate readings of Versailles? To my mind, it is less the grand axis than the network of bosquets and diagonal circulation and alignments that propagate the central experience of the gardens [12-6]. Modulating openings after constrictions, not any axis in isolation, choreographs the revelation of view and manipulation of scale. The mixture of architectural colonnades with naturalistic spaces reinforces the basic rhythm of Le Nôtre's design. This is not to say that the grand axis is

12-5

André le Nôtre. Versailles, France, 1660s+.

The restricted view of the camera lens heightens the importance of the axis as the defining element of the garden.

Marc Treib

12-6

Versailles, France.

The alternate pathways through the woods and clearings of the Petit Parc offer a far richer landscape experience quite in contrast to the principal axis normally featured in photographs of the garden.

Marc Treib

of little consequence; without doubt, it serves as an armature for the full flesh of the park as the human skeletal system serves as the armature for the body. It may also have been Le Nôtre's principal idea. I mean only to point out that our continued use of a particular "frame"—in this case, the axis—greatly biases our reading and understanding of this landscape.

The role played by the photographic frame is perhaps even more decisive in gardens that sought nature as their model, such as those of eighteenth-century England. In many instances, the landscape design was contrived to be seamless, merging the limits of the garden with its wooded or agricultural surroundings. For these gardens the rectangle of the photograph creates entirely new visual relationships, perhaps only suggested by the play between the architecture of the manor or folly and the landscaped grounds.

There seems to have been no escaping the conventions of traditional pictorial representation in photographs of these landscapes. Claude Lorrain's use of the foreground tree, with its overhanging branch, to frame the view and create depth, traces a long history through sketches, paintings, and finally photographs. In the Red Books that served as design presentations, Humphry Repton used rendered depictions that acquired extra power in the "after" images, which tended to be far better composed than the "before" views (see Stephen Daniels on Repton, in Chapter 2 of this volume).

Repton appears to have been quite aware of the power gained by playing the regular against the natural, the smooth against the rough. In discussing architectural style, for example, he demonstrated how one should pit the regularity of the Grecian style against the jagged silhouettes of conifer woods. Conversely, the neo-Gothic demanded the soft forms of deciduous plantings.[11] In a similar way, much contrived naturalism benefits from the photographic frame because the frame limits the field perceived. It also organizes masses of vegetation and creates vistas that might be more difficult to apprehend on site. The work of land artists and sculptors continue this tradition. The photograph has become the primary documentation for works set in the American Southwestern deserts that are almost never seen in the flesh (or in the earth, as it were).[12]

One might also cite the work of Andy Goldsworthy, whose installations in nature are normally more evident and robust in photos than in reality.[13]

The rectangle of the camera itself composes the subject of the sculpture, incising it, and removing it from its greater context. Not only does this reconstruction heighten the sense of presence and composition, it also greatly distorts the sense of scale, rendering the installation more impressive in photographs than within the unbounded expanse of their forest sites.

That the photograph has the force to render similar objects or spaces of immensely different sizes is one of its greatest powers. In the past, in drawings and engravings, the human figure often instilled the view with a sense of scale.[14] Indeed, in the prints of several artists—Giovanni Battista Piranesi being the most notorious—the diminutized size of the figure renders the space larger, if not heroic. This willful manipulation of scale seems to be less prevalent in photographs today, however, perhaps due to the tendency of card-carrying landscape photographers to eliminate people from their images (interestingly, *Sunset* magazine, which proclaims the joys of Western Living, always includes people in their photographs). No intention to deceive, notwithstanding, a garden in an 8" x 10" print is far smaller than it measures in three dimensions.

III. TIME AND PLACE

It is difficult for us today to understand how radically photography has changed our notions of time and place. In terms of time, the photograph extends life long beyond the death of the subject. Prior to the photograph, it was the portrait of human or landscape that afforded this comfort only to the wealthy; the photograph has made eternity available to everybody. In *Camera Lucida*, Roland Barthes refers to the camera as a "clock for seeing."[15] To Barthes, central to photography is the fact that what we see within the photograph is a condition he terms "that-has-been."[16] It was at one time real, even if that reality no longer continues today.

Landscapes are ephemeral—ever-dependent on maintenance—and limited by the lifespans of their vegetation and in some instances, the lifespans of their owners. The photograph captures a single moment of their existence and makes of that instant an eternal present, to borrow a title from the architectural historian, Sigfried Giedion.[17] While a landscape does not move like an automobile or a human in motion, it is a fleeting subject. Change lies at the root of garden making. We can accept change in the garden, designing

for color and mass and vista throughout the seasons; or we may attempt to thwart change through our mastery of natural process—as in the French formal garden or most topiary. The photograph depicts time stilled. As a result, we are less aware of the flow of natural and human process in designed land-scapes, and, at best, we trace the procession of evolution through distinct panels, as in a series of freeze-frame images. These images of Sceaux outside Paris, originally to André Le Nôtre's design, unfold the stages in the garden's dramatic return: from the desolate state of the central canal early in the twentieth century through its replanting with the poplars for which it has become known [12-7, 12-8].

The introduction of color to film added another perceptual dimension to the picture, although it did not fundamentally change our understanding of garden space. On the other hand, there is little denying the power of color, as we can see returning once again to the work of Luis Barragán, in this case, the roof terrace of his own house. Just when the photo is taken, of course, plays a significant role in its conveying of space and form. The shift in vegetal color and volume evident in images of landscapes taken throughout the year reflects more than the seasons in which they were taken: they fundamentally modify our reading of spatial depth and the phenomenological positioning of the view's principal garden elements: the walls, the burial areas, and the trees of the Woodland Cemetery, for example [12-9, 12-10].

The photograph has conquered more than time alone; it has also conquered space and scale, offering a simultaneity of exposure never possible in actual life. In *The Voices of Silence*, André Malraux was the among the very first to cite the studies that photography has made possible.[18] Prior to the photograph, he writes, it would have been impossible to compare, say, two paintings by Goya in separate collections. With one picture in the Louvre in Paris and the other in the Prado in Madrid, two days' travel was necessary to see both paintings. They could never be seen within the same gaze—until the photo-graph captured at least the rudimentary aspects of the painting.

As it allowed this juxtaposition of discrete objects, Malraux believed, so photography also falsified scale. He used two sculptures of different dimen-sions to demonstrate his claim. Despite their varying size, in their respective photographic reproductions, the sculptures give the impression of measuring about the same. The single view common to art publications not only dis-

12-7
André le Nôtre. Sceaux, France, 1680s.

The photograph stops time, capturing the canal and park in decay, c. 1920.

12-8
Sceaux, France.

The park and canal restored; the alignments of poplars matured, 1994. Photographs permit comparisons of time and place.

Marc Treib

torts our reading of scale, but also condenses the complexity of sculpture's three-dimensional form reviewed over time to a single canonical view. Time stilled in the photograph denies the complex mosaic of views seen through time, either as the object rotates before us or we circumambulate around it. In some few publications, such as in the book introducing the sculpture park at Storm King, New York, a minimum of two images portray each sculpture, suggesting the infinite number of regards that remain possible.[19]

We can imagine, then, what is denied us in photographs of gardens and other designed landscapes. What had been a multi-sensual experience, haptic in its relation to the size and senses of our body, is reduced to a flat plane. In that sense, the photograph shares the properties of all categories of images that preceded it in time. But the construction of optical systems has also affected how we perceive the photographic landscape, thus distinguishing mechanical photographic capture from the mediated drawing. The use of wide-angled lens distorts through inclusion; the telephoto lens through restriction and flattening. Perspective correction denies the position of eye level before the subject.

We have all experienced the sense of surprise or loss when visiting a place which seems vastly smaller—or at times vastly larger—than what we had predicted from photos. Perhaps, this is not a bad thing, however. We might even venture that this denial of prediction can heighten the basic experience. Robert Venturi, for example, argued in *Complexity and Contradiction in Architecture*, that it was just this distinction between our projected experience of a structure when viewed from the exterior, and the actual reading of the interior upon entry, that instigated much of architecture's energy.[20] Extending that thinking, we might propose that the further the photograph departs from the actual landscape, the more powerful the impression will be, either as pleasure or disappointment. Perhaps.

The elucidation of the properties of photographs in the preceding paragraphs is not intended to condemn them as inherently deceitful and unworthy of capturing the designed landscape. To the contrary, this discussion is intended to caution both the maker and the reader of the photographic image that, rather, one should seek more than verity from photographs. Instead we should appreciate both the medium's possibilities and its limits to record, reduce, augment, enhance, and even recreate our experience of gardens, parks, plazas, and other forms of landscape architecture.

12-9
Gunnar Asplund and Sigurd Lewerentz. Woodland Cemetery, Stockholm, Sweden, 1915–1940.

Photographs capture the seasons, demonstrating dramatic changes not only of color, but also spatial definition. Summer.

Marc Treib

12-10
Woodland Cemetery, Stockholm, Sweden.

Winter, when vegetal mass turns to arboreal framework.

Marc Treib

IV. DISPLACEMENT

Much has been written in the past decade about the privileging of vision—
or ocularcentrism, to use a more academic term—and that too much emphasis
is placed on the visual sense alone.[21] Due to photography, and its extension
in film, video, television, and electronic diffusion, the image has come to
dominate our experience of place. "The primitive notion of the efficacy of
images presumes that images possess the qualities of real things," writes
Susan Sontag, "but our inclination is to attribute to real things the quality of
images."[22] In the 1960s and later, the Situationists and other French theorists
advanced the Latin word *simulacrum* as a marker for contemporary life.[23]
We dispatch fresh food for its frozen variant; we prefer the film to life; the Las
Vegas image of Old Europe to the place itself. And why do we now propose a
virtual reality rather than improving the one in which we have been living?[24]
That we are let down by actuality may be a bad sign; is there any way that
a real landscape can match in power that single, condensed, and perfected
image such as Salas Portugal's Barragán?

Perhaps the lesson here is to avoid the expectation of completion and perfection. A real landscape offers sensual stimuli far greater than those offered by
granules of silver halide or color dies on paper. There are the other dimensions to apprehend—thermal, acoustic, haptic, and olfactory. With vision,
these senses are the basis of our perception and enjoyment of designed
landscapes. It is said in their first confrontations with photographs Native
Americans feared the images might capture their souls. In some ways, they
may have been correct, although perhaps it is the body that is captured more
often than the soul.

We need keep in mind that a photograph is not the landscape, but a vision
of a landscape captured on light-sensitive material by a human being. It is
thus an act of appropriation that may or may not agree with the vision of
the garden's maker or even the particular qualities of the place as perceived
by any individual. This disjunction, this disparity, between place and image,
can be misleading or even purposively deceptive. But it may also be a creative
act that grants to the landscape greater understanding and appreciation,
not to mention an extension of existence in terms of both time and place.
If anything, however, these thoughts should constitute a call to arms for
creating landscapes that offer us far more than what the photograph can
capture, offerings for the soul as well as the body.[25]

NOTES

1 Susan Sontag, *On Photography*, New York: Farrar, Straus and Giroux, 1977, p. 7.

2 Ibid., p. 11.

3 William Ivins, *Prints and Visual Communication*, Cambridge, MA: MIT Press, 1953/1969.

4 John Berger, *Ways of Seeing*, Harmondsworth: Penguin Books, 1972.

5 Jonathan Crary, *Techniques of the Observer: On Vision and Modernity in the Nineteenth Century*, Cambridge, MA: MIT Press, 1992.

6 Michael Podro, *Depiction*, New Haven, CT: Yale University Press, 1998, p. 17. In addition: "At the very least our interest in how things appear in painting is unlike our nonpictorial experience of them, and the difference is not reducible to the fact that they appear in a painting but how they appear in a way that is distinctive of painting" (ibid., p. 18). The same can be said of photography as a medium.

7 Marc Treib, "Inflected Landscapes," *Places*, 1 (2), 1984; reprinted in Marc Treib, *Settings and Stray Paths: Writings on Landscapes and Gardens*, London: Routledge, 2005, pp. 52–73.

8 According to Sontag, "To photograph is to appropriate the thing photographed. It means putting oneself in a certain relationship to the world that feels like knowledge—and therefore, like power," *On Photography*, p. 4.

9 See Marc Treib, "Aspects of Regionality and the Modern(ist) Garden in California," in Therese O'Malley and Marc Treib (eds), *Regional Garden Design in the United States*, Washington, DC: Dumbarton Oaks, 1995, pp. 5–42.

10 Michael Heizer's *Double Negative* (1970) consists of two aligned bulldozer cuts on the edge of a mesa in Nevada. As one's eye descends along the inclines at either end, the side walls grow in their intensity, directing the view firmly ahead. See Michael Heizer, *Double Negative*, Los Angeles: Museum of Contemporary Art, and New York: Rizzoli, 1991.

11 After discussing the characteristics and relative merits of the Gothic and Grecian (i.e., Classical) styles, Repton concludes: "The outline of a building is never so well seen as when in shadow, and opposed to a brilliant sky; how when it is reflected on the surface of a pool: then the great difference betwixt the Grecian and Gothic character is more peculiarly striking" (Humphry Repton, *Fragments of the Theory and Practice of Landscape Gardening*, London: T. Bentley and Son, 1826, p. 4).

12 The principal documentation for Robert Smithson's land art is found in photographs by Gianfranco Gorgoni; Wolfgang Volz continually photographs the short-lived interventions by Christo and Jean-Claude. The standard work on the subject is John Beardsley, *Earthworks and Beyond: Contemporary Art in the Landscape*, New York: Abbeville Press, 1984.

13 That Goldsworthy is better known in publications than in actual presence is evident from the large number of books on his work. For example, see the early compendium, *Hand to Earth: Andy Goldsworthy Sculpture, 1976–1990*, New York: Harry Abrams, 1990. Since many of the works in the book were temporal and/or performed, the photograph becomes the sole evidence of works that have long since disappeared by the time their images appear in print.

14 See Dianne Harris and David L. Hays, "On the Use and Misuse of Historical Landscape Views," Chapter 1 in this volume.

15 Roland Barthes, *Camera Lucida*, trans. Richard Howard, New York: Hill and Wang, 1981, p. 15.

16 Ibid., p. 77.

17 Sigfried Giedion, *The Eternal Present: The Beginnings of Architecture*, New York: Pantheon Books, 1964.

18 André Malraux, *The Voices of Silence*, Princeton, NJ: Princeton University Press, 1978.

19 John Beardsley, *A Landscape for Modern Sculpture: Storm King Art Center*, New York: Abbeville Press, 1985.

20 Robert Venturi, for example, argued that the detachment between the exterior appearance and the internal spaces of baroque architecture was one source of their impact. *Complexity and Contradiction in Architecture*, New York: Museum of Modern Art, 1966.

21 See, for example, Hal Foster (ed.), *Vision and Visuality*, New York: Dia Art Foundation, and Seattle: Bay Press, 1988, especially p. 11; and Donald M. Lowe, *History of Bourgeois Perception*, Chicago: University of Chicago Press, 1982.

22 Sontag, *On Photography*, p. 158.

23 See Guy Debord, *Society of the Spectacle*, Detroit: Black and Red, 1967/1983, and Jean Baudrillard, *Simulations*, trans. Paul Foss, Paul Patton, and Philip Beitchman, New York: Semiotext(e), 1983.

24 Ada Louise Huxtable discusses the gradient of the genuine and the fake in *Unreal America and Illusion*, New York: W.W. Norton, 1997.

25 Let me cite Susan Sontag one last time: "Nobody ever discovered ugliness through photographs. But many, through photographs, have discovered beauty." *On Photography*, p. 85.

13

Kenneth Helphand

Set and Location:
The Garden and Film

All film takes place somewhere and, especially in the past decade, film scholars have paid particular attention to the landscape of film from diverse points of view. It has been looked at as a subject, most often in the documentary tradition; as a setting, the backdrop and atmosphere of a story; it has been analyzed symbolically; and it has been described as an actual character in the film, an anthropomorphizing of the landscape that interacts with its human characters, with its own personality, its motivations, history, and destiny.[1]

In certain genres, most dramatically the Western, the landscape is central to the film's characterization and identity. The examination of the complex and changing meanings of wilderness and civilization has also been a particular concern. (Although "Westerns" now take place as far afield as outer space, they still partake of the meanings accrued in the landscapes of the American West.) The presentation of the city in film has also been the subject of much discussion. In 1994, the Getty Museum held a month-long symposium on exactly that topic, and there is a burgeoning literature on the subject by film scholars, architects, and cultural historians.[2] For example, architect James Sanders' book, *The Celluloid Skyline: New York and the Movies* (2001) addresses the co-evolution of the building of New York and its portrayal in film.[3] But despite this increasing attention to the landscape in recent scholarship and critique, there has been little explicit discussion of our topic, the designed landscape—although questions of representation are fundamental to any discussion of film, its methods and its meaning. Yet there are films in which garden imagery is fundamental to the director's vision and some of these have become part of the story of both film and garden histories.

Not only has film served as a medium of landscape representation, but also a series of instructive parallels link film theory and technique with landscape design practice. Cinema designers have consciously exploited some of these associations; more often they lie hidden awaiting our awareness. Since the beginnings of motion pictures in 1895, a continued theoretical discourse has aimed at understanding the medium's distinctive characteristics and its relationship to its sister arts of photography, painting, and literature. Erwin Panofsky's insights contained in his classic (1934) essay "Style and Medium in the Motion Pictures" have a particular bearing upon landscape architecture.[4] He writes that film afforded "unique and specific possibilities" which "can be defined as dynamization of space and, accordingly, spatialization of time." This union of space and time is fundamental to the way film portrays any phenomenon because the technology of film modulates space and time. Filmmakers can examine any scale of space or time, and through the devices at their disposal—such as editing and montage—manipulate our resultant impressions.

13-1
Georges Méliès, director.
Voyage to the Moon, 1902.
Star Film

The range of possibilities is endless, but certain conventions are critical. Think of the obvious ones: the aerial view that establishes location and then zooms into a specific site; the tracking shot that simulates the viewpoint of a person walking through a place; a panning shot that sweeps from one side to another, thus capturing a panorama; the static shot that frames the view. Film, unlike still photography, can record the motions of phenomena but also track the moving camera as well. As Panofsky noted, when watching a movie,

> [The spectator] is in permanent motion as his eye identifies itself with the lens of the camera, which permanently shifts in distance and direction. And as movable as the spectator is, as movable is, for the same reason, the space presented to him.[5]

Russian filmmaker Dziga Vertov referred to this as Kino-Eye, the cinematic eye.[6]

To be sure, film records and documents landscape, but it can be understood itself as a landscape, an "illusionary, three-dimensional" world that is the "cinematic landscape." Jeff Hopkins writes:

> This landscape has its own geography, one that situates the spectator in a cinematic place where space and time are compressed and expanded and where societal ideals, mores, values, and roles may be sustained or subverted. The pleasure of film lies partially in its ability to create its own cinematic geography . . . The cinematic landscape is not, consequently, a neutral place of entertainment or an objective documentation or mirror of the "real," but an ideologically charged cultural creation whereby meanings of place and society are made, legitimized, contested, and obscured.[7]

In the cinema, there is always a calibration between the real and reel world, the actual and the screen world.

The first films were created just over a century ago. In the work of three of cinema's innovative pioneers—Georges Méliès and August and Louis Lumière—we can already witness the crystallization of distinct approaches and methods that echo throughout the subsequent century of filmmaking. There is also an uncanny relationship between their work and garden and landscape architectural design. The Lumière brothers attempted to capture reality and present it to an audience: the first use of what became known as *cinema vérité*. When first seeing a train entering a railroad station on film, viewers jumped up in fear and expectation of the locomotive emerging from the screen. However, viewers soon learned that the two-dimensional screen could present four-dimensional phenomena, and, thus, there was no cause for alarm. Two of the Lumières' early films, screened in 1896, included *A Friendly Party in the Garden* with men playing cards and *Boys Sailing Boats in the Tuileries.* In contrast, Georges Méliès' approach derived from the illusionist world of magic and the theater. He created films using elaborate sets that acted as frames for the cinematic action. His constructions were clearly just that, and his intention was not to create a set that obscured its artificiality but instead one that reveled in those qualities. Our knowledge of the process called our attention not only to what lay before the camera, but also to things behind it. His best-known work, *Voyage to the Moon*, featured a magnificently fanciful lunar landscape not only of craters, but also an underground world whose flora were giant mushrooms [13-1].

Do we not see here parallels with the designed landscape? The dialectics of those who would obscure their "hand" and its manipulations to appear as if it was always there as a natural scene versus those who accentuate that which has been created, and even revel in its artificiality. The Lumières' approach is one of discovery while Méliès reveals construction. At first, these might seem opposing poles pitting reality against artificiality, but of course the story is more complex, for each possesses aspects of the other within it. In the contrast between these two approaches, we see distinctions in how the world, and designed landscapes, are represented. But in them we also see parallel approaches to design practice, a set of polarities shared by the film arts and the arts of environmental design.

Among other properties, these two approaches embody the dialectic between set and location. Professional scouts seek out places as preferred locations for films.[8] Scouts use multiple criteria for selecting locales that fit the film's narrative as well as the pragmatics of its shooting, crew, access, and cost. In searching for locations, they respond to the desires of producers and directors for whom verisimilitude is essential, or where the illusion of that locale is possible. Thus, just as actors have stand-ins and body doubles, so do places: Toronto can be New York, Helsinki becomes Moscow. Every state and many cities establish film commissions that court filmmakers in full realization that a production company in the community brings in substantial sums of money, and that the film itself can function as a form of advertising. In 1996, The *Royal Institute of British Architects Journal*, well aware of the promotional character of film, published an article "Location Is Everything," explaining how to get its members' buildings into the movies.[9] Shooting on location concludes a process of discovery. Filming may be taken as form of landscape design: site selection based on a specific set of criteria, and then — like a site design—once selected, the location exerts its own often unforeseen power upon the making of the film.

Films may be shot on location, but most are made in studios. A set designer is now often called "production designer" (PD) or "visual consultant." Set and location establish time and place, as well as mood and atmosphere, and they are coded in film genres. Landscape designs act in similar ways and are equally coded. The set, like the garden, can achieve a kind of hyper-reality, an intensification of its elements and spaces that has been called a "magical artificiality" and "invented reality."[10] The extreme example of this in the art of film is "Caligarism," a term derived from the German film *The Cabinet of Dr. Caligari* (1919) with its dramatic Expressionist sets; here the film invents, or literally constructs, its reality [13-2].

Like the garden, film alludes to other places although rooted in a certain reality. Both arts draw upon a self-referential history. Film sets, unlike theater sets, are designed to accommodate the camera, which can be placed any-where.[11] The camera moves through the set as our surrogate, directing our attention and manipulating our thoughts and emotions. The set is normally seen as a backdrop, background, or more generally as an atmosphere that is rarely foregrounded; at times, however, its presence is so strong that it imposes on the action or narrative. In that sense, the set shares the fate of many designed landscapes which have become, as Walker and Simo have reminded us, "invisible," and thus their full impact is neither felt nor under-stood, for it is assumed to be a natural and expected presence.[12] Only when foregrounded or asserted is its influence, and perhaps design, apparent to the viewer.

Many writers have noted how both set and location can act as cinematic characters. Sets blur the boundaries between the real and the artificial, indoors and out—although most sets are constructed on indoor sound stages.

13-2

Robert Wiene, director.
The Cabinet of Dr. Caligari,
1919.

Decla-Biscop Ag, Germany

Set and location often intersect, as sets are constructed on location. The use of matte devices adds another dimension, where an artwork is created and then filmed in combination with live action. The developing techniques of computer-generated images (CGI) have actuated the complexities of our experience of the illusions presented to us on screen. One can also suggest that the act of shooting itself converts a locale into a set by becoming part of the film. As a result, the act of making any place a "location" highlights its theatrical, scenographic, and cinematic aspects.

There is a parallel between reality-versus-illusion exemplified by the Lumières and Méliès, and the dialectic of set and location—a situation augmented by the framing of the lens. The framed cinematic image may be either open or closed. The open frame acts more as a window into the scene, while the closed frame acts more as a picture frame containing the image within. (And of course these polarities are just that, with a multitude of gradations in between.) As Leo Braudy notes, "The director of the closed film has therefore created his own space, while the director of the open film has found space within which to tell his story."[13] Once again, set and location. The filmmaker, Jean Renoir, the son of the celebrated painter, pioneered the use of the open frame and used lateral camera movements to capture a continuous reality. "The blackness surrounding the screen masked off rather than framed the image."[14] The Soviet filmmaker, Sergei Eisenstein,

and the psychologist, Rudolph Arnheim, emphasized that the spectator sat before a framed image.[15] In fact, this relationship extended the Albertian Renaissance concept of painting in which the perspectival image was conceived as a representation of the world viewed through an open window. André Bazin, in fact, likened the screen to a window, and the analogy of the window abounds in film theory.[16] These all reflect film's debt to the worlds of painting and theater.

One last concept requires mention in this context. The *mise en scène*, which literally means "put in the scene," refers to the totality of the world created within a film through all the devices available to the filmmaker: framing, composition, lighting, sound, set, and action. Here yet another parallel connects film space and garden space. Films always frame what they seek to portray, and therefore the crafting of the *mise en scène* is the design of a cinematic space. Garden designers display a similar dexterity in the making of spaces that serve to highlight a site's narrative. Other relationships are numerous. The close-up shot has a particular resonance for landscape design as we oscillate between the field of view and focused attention. Design encourages the close-up: for landscape designers details are its physical equivalent.

The specific films cited here to illuminate these ideas by no means constitute an exhaustive list, but represent films often cited in discussions of cinema. Thus, these works possess a certain significance in the sense of having become part of our "literature." One common approach to film studies centers on the study of type and genre. Of these, I will focus only one genre, however: the period-costume drama.

There are, of course, documentaries that address the designed landscape.[17] Documentaries record a place and take us there through the eyes of the filmmaker, although they result, of course, from collaborations between a writer and, at times, a host and sponsor. Typically, documentary relies on certain conventions: authoritative or engaging voiceovers, talking heads, scores of period music, lingering close-ups, slow pans, the occasional use of historical materials such as paintings or photographs—but rarely plans. People are excised from the scene or for a contemporary site, they provide elements that portray a vivid slice of life. Most garden films, however, resemble garden magazines with a horticultural and photogenic emphasis. They have much in common with the travelogue, enticing us to visit on our vacation—which need not be taken as a disparaging comment.

Without question, the documentary constitutes an important form of cinema, but instead I will focus on films that are typically characterized as costume dramas and/or period pieces. The medium of film contemporizes, making the past present in the dual spatial and temporal meanings of the term—feeling both here and now. This is especially relevant for the world before the twentieth century, a world that was never actually filmed, although it had been photographed for some seventy years. Thus, for any era before 1895 the film is a form of reconstruction, restoration, or reenactment, a property explicit in films involved with historical periods. In the costume drama, the set or location itself is one of the "costumes" by which the place is adorned, much like the proper wardrobe worn by the actors. Throughout the following films the relationships between garden and costume are made vivid, bringing to the fore the rarely examined connection between garden and fashion design. These films restore a time and place.

Roberto Rossellini's (1956) *The Rise of Louis XIV* appears documentary in style, a bit of the "you are there" school where we feel ourselves to be witnesses to the events of the past—it may even possess a didactic role as well, with its dialogues providing a crash course in French history.[18] The location scenes of the young king and his court at Versailles are striking, for while one could compose such a scene in a photograph or in a "garden" film, we normally do not. Thus, in the film we view a historical reconstruction that provides us with a glimpse into at least one small portion of the vast royal garden, and the ensemble of costumed courtiers, fountains, and plantings. The figures of the king and the members of his court, as much as precisely pruned vegetation, are all cast as components of the garden's parterres. The action then enlivens these theatrical metaphors for the French garden, vividly showcasing its play, actors, and stage.

Essential to Stanley Kubrick's (1975) *Barry Lyndon*—an adaptation of William M. Thackeray's novel—was a very precise spatial, cultural, and temporal reconstruction of the book. Portions of the film transpire at an eighteenth-century pace, or at least so they appear: certainly the pacing is far more languid than that to which we have become accustomed. The camera lingers; often, there is no conversation. The film also underscores that all film contemporizes. The fictional Castle Hackton is a conflation of the English gardens at Castle Howard and Stourhead, although neither site is identified in the film [13-3]. What we do witness are places brought to life by the exploits of the actors, as the film's garden scenes represent the upper-class life of the era. The film presents particularly well the ephemeral events that occur within a garden's more permanent frame—events such as theatrical performances on temporary stages, games, fishing, and tents erected on the lawn. Kubrick carefully situates the action—events happen just where the story suggests or demands. He slowly zooms in or out, thus situating the characters in space. He has an eye for accuracy, although his homage to painted scenery is also clear. He even ambitiously attempts to approximate the era's sensory environment: the film was noted for its interior scenes shot by true candlelight. Importantly the garden is placed in context—physically and socially—as one of a set of landscapes that run the gamut from palace to battlefield, road to riverbank.

Last Year at Marienbad, the 1961 film by Alain Resnais, based on a screenplay by Alain Robbe-Grillet, was filmed at the Nymphenburg gardens in Munich and presents cinema's perhaps most recognizable garden imagery. Although the film was shot on location, the garden clearly functions more as a set than as a vital landscape. The scenes oscillate between interiors of richly gilded corridors—baroque and seemingly endless—of a hotel spa whose guests are always formally attired. Throughout the film, mirrors abound, as do trompe l'oeil garden images and even a framed plan of the gardens.

The grand axial view of the garden is returned to throughout the film. Robbe-Grillet was explicit in his directions. His screenplay calls for a garden *à la française* with a minimum of vegetation except shrubs perfectly clipped, with huge statues and a landscape that is essentially empty, "without a single living being." In addition, he specified slow lateral views showing perspectival views of paths, cones arranged in rows, and clipped hedges. Resnais's realized film matches in care, composition, and framing the precision of Robbe-Grillet's directives. The meaning is enigmatic, yet the unnamed couple who meet at the Marienbad resort inhabit a "petrified garden," a "garden carved in stone"—a reference both to a cemetery and to the frozen state of the characters. The shots compare statues and people, spatial structure and emotional relationships [13-4]. The figures resemble statues and topiary, static and inert. In the final lines of the film, one of the protagonists declares:

13-3
Stanley Kubrick, director.
Barry Lyndon, 1975.
Warner Brothers

13-4
Alain Resnais, director.
Last Year at Marienbad, 1961.
Wellspring, formerly Winstar Cinema

the park of this hotel was a kind of garden *à la française* without any trees or flowers, without any foliage...Gravel, stone, marble and straight lines marked out rigid spaces, surfaces without mystery. It seemed at first glance, impossible to get lost here...at first glance... down straight paths, between the statues with frozen gestures and the granite slabs, where you were now already getting lost, forever, in the calm night, alone with me.[19]

There is a gap between "at first glance" and the slowly revealed but never-completely understood mysteries hidden within the tale and implicitly within the garden itself. Perhaps the film's most striking shot is that of stationary figures standing on the garden's main axis. The sun is bright and their shadows are long, but they are the only shadows in the scene; none are cast by the plants!

Peter Greenaway's (1982) *The Draughtsman's Contract* owes a considerable debt to *Last Year at Marienbad*. The methods, meaning, and pitfalls of representation are all explicit themes in this film, which takes place at the country estate of Compton-Anstey in the year 1694. The artist, Mr. Neville, is contracted to produce twelve exterior views of the house and garden in the same number of days (actually an excellent drawing assignment) [13-5]. The choice of views is left to the draughtsman and the film carefully enumerates each site, many of which we return to several times as the drawing and the story progress in carefully contrived layers [13-6, 13-7]. We see Neville first choosing a site and then setting up his painting equipment, his chair and easel, and, most importantly, the gridded frames through which he will view the landscape. This device enables him to carefully transfer the grid in the frame to the grid on the paper to accurately construct the perspective drawings.

Throughout the film, we bear witness to his technique and the power of the frame—of course, watching the film we are looking through yet another frame. Greenaway's cinematic frames are carefully composed and closed, while the draughtsman Neville's frames are depicted as open although the artist strives to construct what occurs within them. Neville demands certain conditions of his clients: the areas are to be kept clear; servants are not to work in the area under study; no one is to enter the scene. These are to be peaceful and quiet garden views. He is not dogmatic, for in Drawing Five, the critical hilltop prospect of the estate, he says that "such animals as are

13-5
Peter Greenaway, director.
The Draughtsman's Contract,
1982.

*Wellspring, formerly Winstar
Cinema*

13-6
The Draughtsman's Contract.

*Wellspring, formerly Winstar
Cinema*

presently grazing in the field will be permitted to do so." The draughtsman has "the god-like power of emptying the landscapes," says his nemesis in the film. He has obvious preferences, the time of day is carefully chosen and he prefers clear skies and sharp shadows. The two great genres of painting in this period, landscapes and portraits, blur, for these are landscapes in which house and garden sit for their portrait. The draughtsman carefully poses elements within the view, adjusting the positions of objects and even the figures where desirable. A portrait of the son-in-law of the owner is taken, but unbeknownst to him he is only the stand-in for the owner whose face will be filled in later. Neither the frame nor the drawings nor the film can be completely controlled. However, one marvelous scene shows sheep in a meadow charging the artist and bursting through the borders of the frame [13-8]. The drawings build, the grid is filled in, the outlines are drawn; shading and nuance are added as we conflate the real subject and its representation.

The Draughtsman's Contract addresses the many roles played by representation and the power, in this instance, that the drawing yields. Neville says, "I hold the delight or despondency of a man of property by putting the house in shadow or sunlight." Most obvious is the drawing's functioning as a record of possession and power. In this instance, the drawings have been commissioned as gift to the owner of the estate and household, or so we think. Rendered in ink because they are contracts—legal documents—the drawings say: it is all here "in black and white." The drawings also serve as clues as the film is ultimately a drawing-room mystery. The drawings are witnesses and are thought to contain clues to the murder perhaps unknown to the artist who executes, but is unable to interpret, his own handiwork. At one point, the draughtsman is told "an intelligent man will know more about what he is drawing than he will see."

Near the end of the film, a Dutch landscape gardener is commissioned to soften the geometry of the garden, construct a lake, and "introduce new ease and complexion" to the garden. The lady of the house notes that "[It is] you, Mr. Neville who opened our eyes to the possibilities of our landscape." Interviewed about the film, Greenaway noted:

> The facets of the drawing and the landscape are compared on another level of representation, the film. I want those three ideas to be present in the whole structure of the movie, so that one is aware that we are making comparisons all the time between the real landscape, Mr. Neville's image of it and ultimately, us as viewers seeing those ideas represented on film.[20]

13-7
Peter Greenaway, director.
The Draughtsman's Contract,
1982.

Wellspring, formerly Winstar Cinema

13-8
The Draughtsman's Contract.

Wellspring, formerly Winstar Cinema

13-9
Sally Potter, director.
Orlando, 1992.
The Sales Co., London

13-10
Orlando.
The Sales Co., London

13-11
Orlando.
The Sales Co., London

Orlando (1992), directed by Sally Potter and based upon the Virginia Woolf novel, is in part the story of a great house and its garden. Woolf's novel dealt with her relationship with Vita Sackville-West and the house in the story is inspired by Knole, the great estate in Kent that was Sackville-West's childhood home—and that also inspired Sackville-West's epic poem "Land."[21] Orlando traces the house, the family seat, and characters over almost four centuries beginning in the year 1600, in seven cinematic chapters: 1600, 1610, 1650, 1700, 1750, 1850, and the time of filming, 1992. The house and garden setting remains constant, thus giving us the rare opportunity to witness the evolution of a landscape over time. The garden location sets the frame while encapsulating each era of the saga. The reconstructions are careful and rich in detail, from objects such as the queen's footstool to social phenomena such as tulipmania. In 1750, the garden is shown being pruned, a rare inclusion of garden labors [13-9]. In 1850, these same plants appear as topiary teacups [13-10]. The fact that in each chapter the characters' garments change, while the setting remains constant, accentuates the relationship between the costuming of people and the look of the place. In the final segment Orlando enters the garden's maze but emerges in the contemporary garden.[22] The final scenes include a fantastic image of trees wrapped in the manner of a Christo installation—trees as outdoor cloth-covered versions of the interior furniture seen earlier in the film [13-11].

One could also cite the Merchant/Ivory productions of novels by E.M. Forster and Henry James, and the over 150 films that have used New York's Central Park as a location.[23] Instead, let us jump in time to *Mon Oncle* (1956) by Jacques Tati. Here the garden is both an actor in the film and an ironic subject. Created in the mid-1950s, *Mon Oncle* today reads as a post-war period-costume drama. Tati satirizes the modern house and garden: the obsession with technology, the forms created at the expense of comfort, the rejection of the past, and modern design as a status symbol. The garden plan is itself a pastiche, with fragments of paths and stepping stones, a patio and fountain —all either too large or too small for the space. A running joke involves a fountain centering on a metal fish that the lady-of-the-house turns on and off depending on the status of whoever passes through the gate. In one scene a guest carefully negotiates the garden's stepping stones as if those on the celebrated pathways of the Katsura villa in Kyoto [13-12].

13-12

Jacques Tati, director.
Mon Oncle, 1958.
Continental Distributing

13-13

Peter Weir, director.
The Truman Show, 1998.
Paramount Pictures

In recent films the ambiguous relationship between set and location remains dramatically present. *The Truman Show* (1998) was shot at Seaside, Florida —a stand-in for the film's fictional town of Seahaven—although few members of the movie audience are aware of that fact [13-13].[24] The film presents a confusing set of conflicting relationships. We, the audience, are watching a film that purports to document a television show, which is seemingly an ultimate "reality show" tracing the entire life of Truman Burbank. But, in reality nothing is real, "it's a set, it's a show," with everyone, except Truman himself, an actor and cast member. Although the set(ting) is the iconic New-Urbanist town of Seaside, the setting owes an equal debt to Disneyland. This is life and community as a theme park complete with a "backstage," the Disney term for everything that is hidden—but essential—to Disneyland's functioning although these are areas never seen by visitors.

In Seahaven, everything is tidy and manicured, the whole place, people included, meticulously coiffed. Cars move slowly; bicycles abound; strangers nod and say "hi" to one another. The film depicts Truman, and the television show, at age thirty. It is a yuppie fantasy, featuring Truman's comfortable home, attractive wife, job, and material rewards, in "a nice place to live," as the town's license plates say. It is a truly adult world with hardly two children in the entire film. The viewers' emotions—both of those who watch *The Truman Show* on television in the film and the film's audience—are ambivalent. We root for Truman to escape this horrific hoax played upon another human being, but recognize that the world "outside" Seahaven is far messier and more complex than the town's perfect world. Is it just a place to visit, like Seaside itself, or does it offer conditions we wish were part of our own daily life?

An unsigned article, "Landscapes from the Screen," in the April 1937 issue of *Landscape Architecture* magazine noted the power of the cinema and hoped for a time when "landscape architects of national reputation will be retained by the great motion picture studios, in the same way that the great producers retain experts in their fields, to endure good taste and truthfulness in the landscape presentation in their picture."[25] The statement is naïve. Nevertheless, it is easy to imagine landscape architects contributing to films not only for their input to historical accuracy, but also for our ways of seeing and representing illusions and reality, set and location.

1 Kenneth Helphand, "Landscape Films," *Landscape Journal*, 5(1), 1986: 1–8 and "Battlefields & Dreamfields: The Landscape of Recent American Film," *Oregon Humanities*, 1990: 18–21.

2 See Mark Lamster (ed.), *Architecture and Film*, New York: Princeton Architectural Press, 2000; Michael Webb, "The City in Film," *Design Quarterly*, 1987; Scott MacDonald, *The Garden in the Machine: A Field Guide to Independent Films about Place*, Berkeley, CA: University of California Press, 2001; Dietrich Neumann, *Film Architecture: From Metropolis to Blade Runner*, Munich: Prestal-Verlag, 1996.

3 James Sanders, *The Celluloid Skyline: New York and the Movies*, New York: Alfred A. Knopf, 2001.

4 Erwin Panofsky "Style and Medium in the Motion Pictures," in Daniel Talbot (ed.), *Film: An Anthology*, Berkeley, CA: University of California Press, 1934/1967.

5 Ibid., p. 19.

6 See Dziga Vertov, *Kino-Eye: The Writings of Dziga Vertov*, ed. Annette Michelson, Berkeley, CA: University of California Press, 1984, and Vertov's classic film *Man with a Movie Camera* (1929).

7 Jeff Hopkins, "A Mapping of Cinematic Places: Icons, Ideology, and the Power of (Mis)representation," in Stuart C. Aitken (ed.), *Place, Power, Situation, and Spectacle: A Geography of Film*, Lanham, MD: Rowman & Littlefield, 1994, p. 47.

8 A 1999 issue of *Outside* magazine highlighted "The 25 Best Careers in the Outdoors." On the same page, location scout and landscape architect are highlighted.

9 "Location Is Everything," *The Royal Institute of British Architects Journal*, March 1996: 30–31.

10 On film and magic, see Stanley Cavell, *The World Viewed*, New York: Viking, 1971.

11 See Leon Barsaqc, *A History of Set Design*, Boston: New York Graphic Society, 1976; Brian Henderson, "Notes on Set Design and Cinema," *Film Quarterly*, 42(1): 1988: 17–28; Bart Mills, "The Brave New Worlds of Production Design," *American Film*, 7(4), 1982: 40–46.

12 Peter Walker and Melanie Simo, *Invisible Gardens: The Search for Modernism in the American Landscape*, Cambridge, MA: MIT Press, 1996.

13 Leo Braudy, *The World in Frame*, Garden City, NY: Anchor Books, 1977, p. 48.

14 Peter Wollen, *Signs and Meaning in the Cinema*, Bloomington, IN: Indiana University Press, 1969. These themes are much discussed in film theory.

15 Sergei Eisenstein, *Film Form and Film Sense*, Cleveland, OH: Meridian, 1942/1949/1957; Rudolph Arnheim, *Film as Art*, Berkeley, CA: University of California Press, 1933/1967.

16 André Bazin, *What is Cinema?* Berkeley, CA: University of California Press, 1967.

17 For example, see *Nature Perfected* by William Howard Adams, directed by Walter Lassally (1995).

18 *You Are There* was a 1950s television show that reenacted historical events.

19 Alain Robbe-Grillet, *Last Year at Marienbad*, New York: Grove Press, 1962, p. 165.

20 Robert Brown, "Greenaway's Contract," *Sight and Sound*, 51(1), 1982: 34–36. See also Amy Lawrence, *The Films of Peter Greenaway*, Cambridge: Cambridge University Press, 1997.

21 Knole is now a National Trust Property.

22 Mazes are also found in Stanley Kubrick's *The Shining* (1980) and Joseph Mankiewicz's *Sleuth* (1972).

23 The films of James Ivory and Ismail Merchant are distinguished by their meticulous attention to period detail in costume and setting.

24 See Fred Bernstein, "Seaside on Celluloid," *Blueprint*, 154 (1998): 42–44; Luis Fernández-Galiano, "El mundo de Truman," *AV Monographs*, 75–76 (1999): 146–147.

25 "Landscapes from the Screen," *Landscape Architecture*, April (1937): 145.

FILMOGRAPHY

Barry Lyndon (1975),
Stanley Kubrick

Boys Sailing Boats in the Tuileries (1896)
Lumière Brothers

The Cabinet of Dr. Caligari (1919),
Robert Wiene

The Draughtsman's Contract (1982)
Peter Greenaway

A Friendly Party in the Garden (1896)
Lumière Brothers

Last Year at Marienbad (1961)
Alain Resnais

Mon Oncle (1956)
Jacques Tati

Nature Perfected (1995)
Michael Gill and Walter Lassally

Orlando (1992)
Sally Potter

The Rise of Louis XIV (1956)
Roberto Rossellini

The Truman Show (1998)
Peter Weir

Voyage to the Moon (1902)
George Méliès

14

Noël van Dooren

From Chalk to CAD:
Drawing Materials in the
Work of Alle Hosper

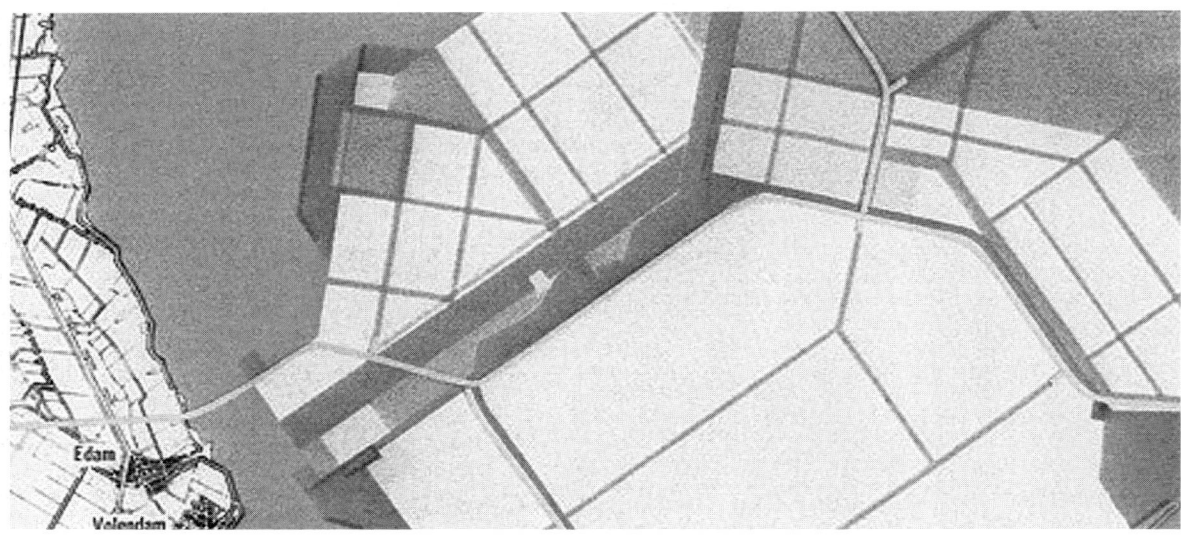

The Dutch landscape architect Alle Hosper died, far too young, in 1997, abruptly truncating a dynamic career that had taken him to all the corners of his profession. A book about this important representative of contemporary Dutch landscape architecture focused on his own professional development within the framework of the development of landscape architecture as a whole.[1] While working on this book, I noticed that his career, which started about 1967, spanned an era in which drawing materials and techniques changed profoundly. Too little research has been done on the question of when and why new materials came into use. Can you still remember the first time you used a color marker? But even less thought has been given to the influence that new materials have on the products of landscape architecture. I tried to discover how the projects carried out by Alle Hosper and his colleagues were related to the means of drawing, as a contribution to a future, more thorough, investigation of the significance of drawing techniques.

14-1

Alle Hosper with others. The Markerwaard, The Netherlands, 1983. Plan.

An early color copy. Hosper and Baljon thought the somewhat blurred nature of the reproduction was appropriate. This study was meant to force a decision—unsuccessfully, as it turned out—for constructing the polder.

A color photocopy: now almost old-fashioned in terms of reproduction technique. In 1983, Alle Hosper and Lodewijk Baljon produced a plan for the much debated new Markerwaard, meant as the final stage of the Ijsselmeer polders project. After much thought about how to present the plan for this polder, they finally decided to insert a color copy in the booklet [14-1]. It was made on the only color copying machine in The Netherlands at that time—at Schiphol Airport. They deliberately chose this particular medium—in those days rather advanced—because, as Baljon stated:

> We used self-adhesive film, without any lines in black ink, to produce a fresh, modern plan. By copying it, we could make the edges precise, but at the same time a bit blurred, so that the drawing did not appear too technical.[2]

Hosper used the computer for the first time in 1992 for a project for an installation for temporary gas storage. The machine in question used computer-aided design (CAD) software that allowed corrections to be made more easily [14-2]. However, it soon became obvious that the computer was also suitable for other activities in the design process. Barely ten years after its initial use for drawing, the computer was being used for almost all presentations. And today, even if initially executed by hand, a drawing is often scanned, reworked, and then fitted into a layout using a computer.

THE PRECISION OF CHALK

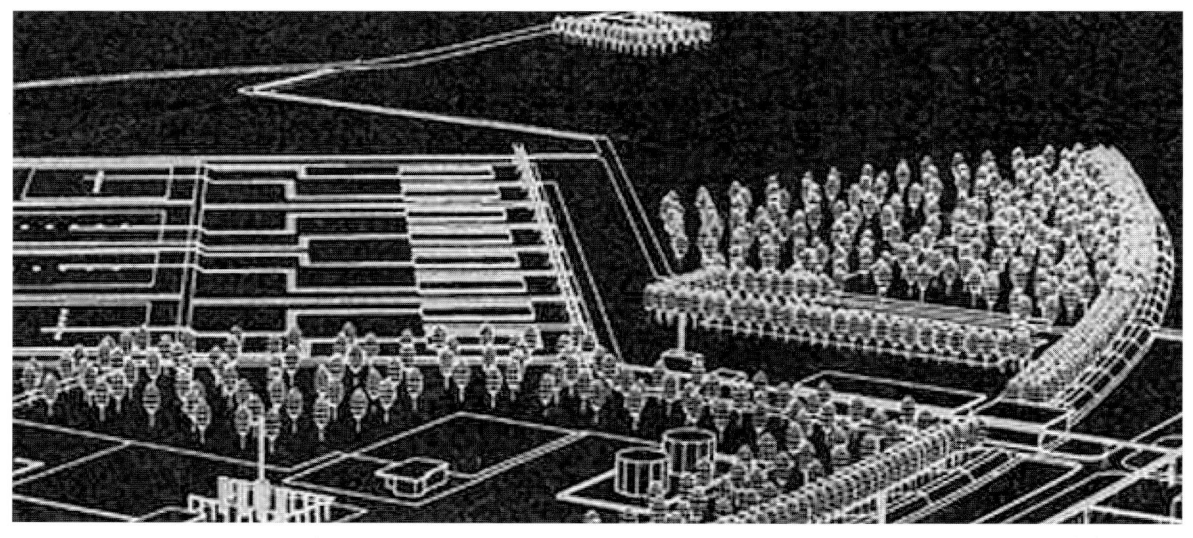

Alle Hosper trained at the landscape architecture department of Wageningen University under the leadership of Professor Jan Bijhouwer, who was a devoted maker of pen and pencil sketches. Hosper was nevertheless seized with enthusiasm for the new possibilities offered by chalk during his work practice at the Staatsbosbeheer, the Dutch Forestry Agency, in 1967. It was easy for Hosper to be inspired by chalk, as he was guided at the agency by the landscape architect Nico de Jonge. Although the use of chalk was not widespread, De Jonge stuck with it for almost all his professional life—and it even characterized his way of working. Chalk was only suitable for making rough sketches. In those days, it was usual at the Staatsbosbeheer to draft the main outlines of the drawing in charcoal and then color and finish it with chalk. In fact, wax crayon was first used, but later chalk was employed, due to its simplicity. While the former is easier to apply, it does require a fixative, and many landscape architects will no doubt never forget the penetrating smell of the spray fixative common in that era.

De Jonge used his chalk technique throughout the design process. Considering the current attention to detail in designing, chalk does not appear to be a very practical medium: it produces thick lines and imprecise borders, and the designer thus is unable to render details to a sufficient level of precision. Ellen Brandes, De Jonge's right-hand woman, turned this assumption around, transforming a liability into an asset. With chalk, she noted, it is impossible to be finicky at too early a stage of the design.[3] Instead, volume and the broad-brush approach become the design's salient features. It is the broad outlines that matter here, and not the width of a grassed roadway shoulder, the planting pattern for an avenue, or the selection of a particular species of plant [14-3].

De Jonge's choice of chalk as a drawing medium was no coincidence. It had an ideological basis also reflected in his choice of colors: blue represents water; orange-brown stands for towns; dirty yellow depicts the remainder. Woodland was always drawn with black chalk. De Jonge stated that green is "a ranger color," referring to the green of the Dutch ranger uniforms. This was not meant as a compliment. Instead, it expressed his irritation at the conservative treatment of the natural environment.[4] Behind De Jonge's typically individualized approach lay an important principle, namely, that color and graphic representations do not need to appear natural—a woodland does not need to be green because trees themselves are green—but rather it is the abstract idea that must be conveyed. De Jonge regarded woodland as a volume. Using black chalk for woodland gives it the necessary clarity and force.

Hosper, having learned the chalk technique from Brandes and De Jonge,

14-2
Alle Hosper. Gas storage facility, Grijpskerk, The Netherlands, 1994. Aerial perspective.

The first drawing for which Hosper's own office used computer-aided design.

Bureau Alle Hosper

14-3
Nico De Jonge. Walcheren, The Netherlands. Landscape plan. Detail of the area around Middelburg.

The drawing shows De Jonge's characteristic rough drawing style with its focus on broad outlines.

used it faithfully for many years. That as a student he already knew the relevance of the chosen drawing technique was revealed in the summary of his professional practice compiled when returning to university. In this report, he wrote: "The way in which materials and colors are used by De Jonge cause his plans to demand attention in a very special way."[5]

NOTATION TECHNIQUES

In 1969, Alle Hosper joined the new Department of Spatial Planning at Wageningen University. There he developed a diagram the department referred to as the "man/place" diagram. This diagram represented the young department's emerging definitions of, and illustrations of, the influence of "place" on "man" and vice versa [14-4]. "We discussed such a lot of things in your office," wrote Hosper in a *Festschrift* presented to Wim van Mourik when he retired as Professor of Spatial Planning at Wageningen. They had mainly talked about the theory of this new discipline in which the relationship between "man" and "place" formed the core. Hosper produced many diagrams to illustrate this relationship. The graphic aspects of these diagrams are remarkable, almost technical in character and without the casual feel of the chalk drawings Hosper had made in the previous year. However, the "man/place" diagram did not stand alone. While working part-time for the department, he also joined the Kring Midden-Utrecht, a group of communities working together on planning issues in and around the city of Utrecht. The Kring Midden-Utrecht made a conscious decision to undertake research into new notation techniques. Hosper was merely one part of a large entity and was unlikely to have had much influence on the group's decision making. However, he was able to determine his own drawing style after completing his studies and went to work for the second time at the Staatsbosbeheer. The diagrams on which the landscape plans for Ameland and Terschelling were based show that the "man/place" diagrams were not just a passing phase in his drawing technique [14-5].

Both the meaning of these diagrams and the techniques used to make them accorded with the environment in which Hosper was then operating. Given the new emphasis on research and scientific analysis at the time, landscape architects could no longer "just" produce designs, as was more or less the case in the work of Nico de Jonge. Each plan had to possess a solid, proven basis. The method of drawing reflecting this more analytical approach was very much influenced by Meto Vroom, who succeeded Bijhouwer as Professor of Landscape Architecture at Wageningen in 1966. However, other disciplines such as soil science, ecology, and sociology contributed equally to the analytical approach. These disciplines also required an innovative way of

14-4

Alle Hosper. "Man/place" diagram.

14-5

Alle Hosper. Terschelling Landscape Structure Plan, c. 1994. Study drawing.

This analytical drawing is part of a larger series in which data were recorded in squares measuring one kilometer on a side. The data were partly specific (for example, the length of the bicycle paths) and partly inferred (for example, the degree of differentiation).

14-6

"Variatie visuele kompleksiteit" (Variations in Visual Complexity), 1974.

An example of the final cartographical analyses from a landscape study of the surroundings of the Dutch city of Helmond.

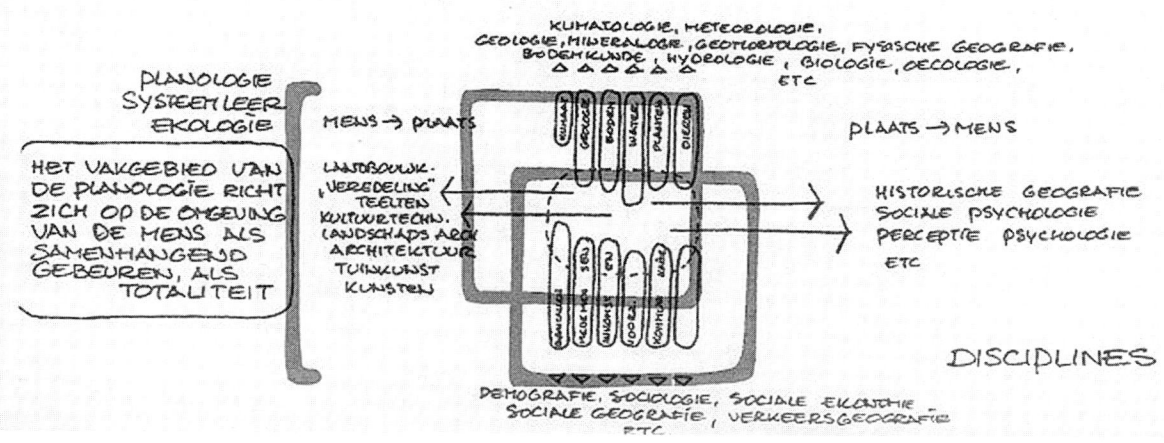

PLANOLOGIE
SYSTEEM LEER
EKOLOGIE

HET VAKGEBIED VAN
DE PLANOLOGIE RICHT
ZICH OP DE OMGEVING
VAN DE MENS ALS
SAMENHANGEND
GEBEUREN, ALS
TOTALITEIT

MENS → PLAATS

LANDBOUW.
"VEREDELING"
TEELTEN
KULTUURTECHN.
LANDSCHAPS ARCH
ARCHITEKTUUR
TUINKUNST
KUNSTEN

KLIMATOLOGIE, METEOROLOGIE,
GEOLOGIE, MINERALOGIE, GEOMORFOLOGIE, FYSISCHE GEOGRAFIE,
BODEMKUNDE, HYDROLOGIE, BIOLOGIE, OECOLOGIE,
ETC

PLAATS → MENS

HISTORISCHE GEOGRAFIE
SOCIALE PSYCHOLOGIE
PERCEPTIE PSYCHOLOGIE
ETC

DISCIPLINES

DEMOGRAFIE, SOCIOLOGIE, SOCIALE EKONOMIE
SOCIALE GEOGRAFIE, VERKEERSGEOGRAFIE
ETC

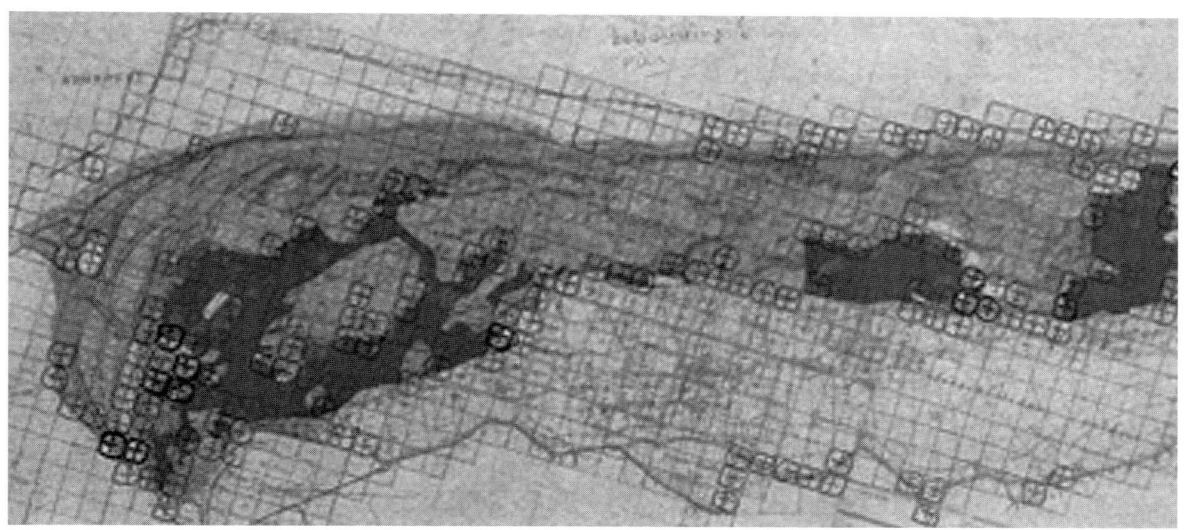

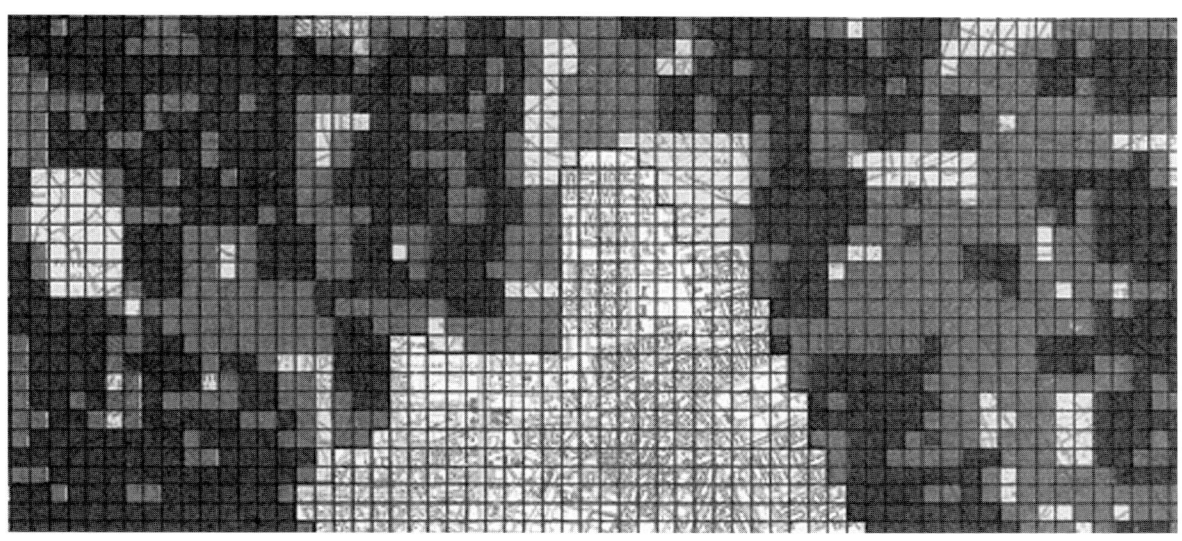

drawing to represent the new arguments they presented [14-6].

In the early 1970s, a new drawing implement, the felt-tip pen, was introduced into The Netherlands from the United States. After only a very short time, landscape architects had enthusiastically adopted this new and versatile tool. Adhesive films took longer to become established; they were expensive and not everyone possessed the necessary skills for cutting and positioning them on drawings. In 1975, Hosper started working at the government department responsible for developing the Ijsselmeer polders. The work pressure there was not that of a commercial firm, so Hosper could execute the drawings himself at a more leisurely pace while learning the ins and outs of adhesive films. When he joined "The Netherlands by Design" in 1985—a spontaneous initiative to brainstorm with colleagues the long-term future of The Netherlands—the ability to use films helped him to take the lead in making graphic presentations. Hosper dedicated many hours to carefully cutting and positioning films.

PRESENTATION

Carefully prepared final drawings had a relatively low status in both the Staatsbosbeheer and Ijsselmeer polders departments. Work had to be executed quickly and the decision-making processes did not require beautiful drawings. "Quite the contrary," joked a colleague of Hosper's, because it was the speed of the tree planting machines that determined the rate at which the plans were made; huge tracts of woodland were being developed at the time. Generally, Hosper and his team only had the chance to make a quick sketch—and the work had to be carried out using that sketch.[6]

In 1985, Hosper moved from a governmental office to a commercial firm:

14-7

Bakker en Bleeker. "Concours International du Parc de la Villette," 1982. Competition entry.

This detail shows the graphically clear, neutrally colored drawings the office "invented."

14-8

Bakker en Bleeker." Stad aan de Stroom" (City on the River), Antwerp, Belgium, 1990. Competition entry.

The theme is the reconstruction of the city's harbor. Both design and drawing style generated heated discussion.

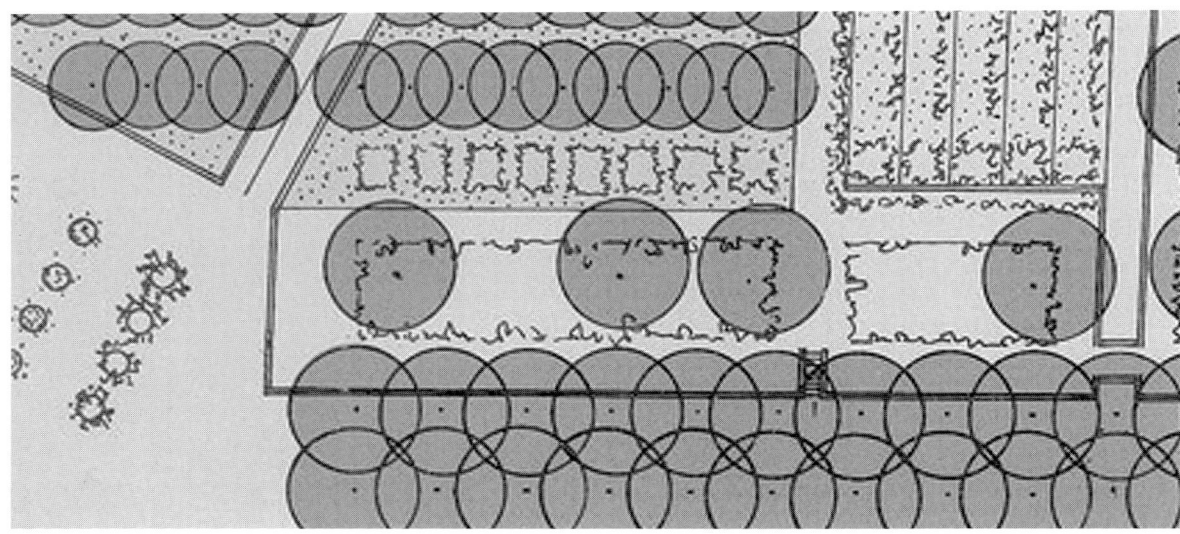

Bureau Bakker en Bleeker (today known as B+B landscape architects). The situation at Bakker and Bleeker was radically different from that in the public sphere and this move had far-reaching consequences for the status and role of certain drawings and the techniques used to produce them. In almost all cases, Bakker and Bleeker worked with a clear program, and in response, the clients received a well-presented product. The office also had a second mission. The firm's principals, Riek Bakker and Ank Bleeker, wanted their firm to stand apart from other members of the landscape profession; not only as far as the content of the design was concerned, but also in the way in which their documents were produced. Their drawings always had a rather business-like look, which Bakker and Bleeker especially liked: graphically clear, neutrally colored, with a sort of emptiness that represented the office's prevailing design ethic: a sober but confident spatial configuration. Their 1982 competition entry for the Parc de la Villette competition displayed all these hallmarks, as well as sufficient graphic beauty to attract the jury's attention [14-7].

In most cases, the method of drawing was determined by the project rather than the person. As a result, Alle Hosper's own style of drawing was not always recognizable in the projects in which he was involved. Despite the prevailing seriousness in their projects, Bakker and Bleeker's landscape designs at times displayed another tendency, one that was more light-hearted. It was Hosper who propagated the more restrained and neutral approach while some of his colleagues followed a more frivolous line. The firm's submission for "Stad aan de Stroom" (City on the River), a competition for the harbor area of Antwerp in 1990, led to extensive discussions on both the nature of the design and its graphic representation [14-8]. The design, for

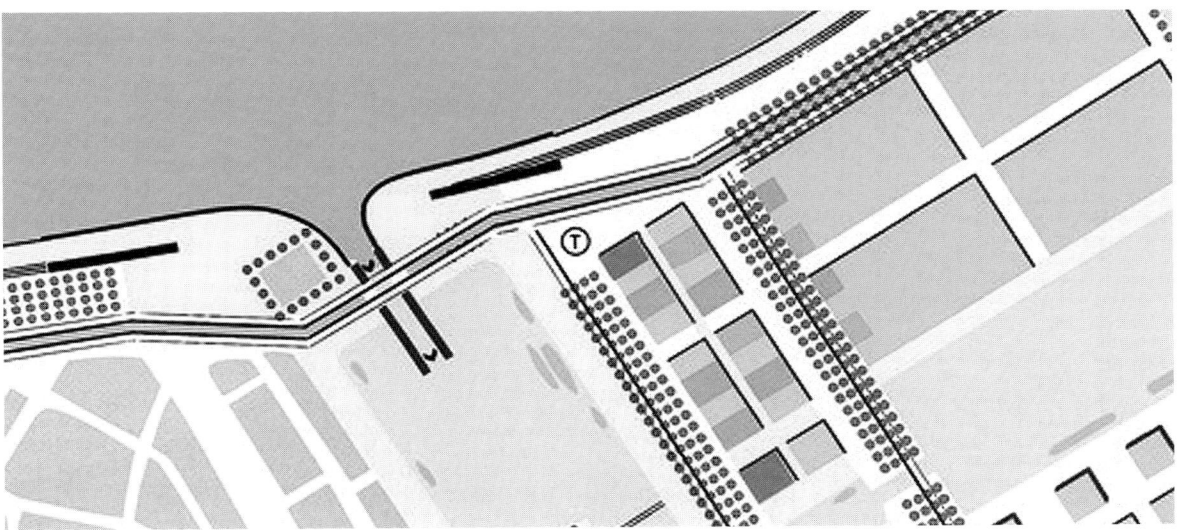

which Hosper took the lead, was deemed too sober by his colleagues. It is worth noting that, in subsequent years, projects designed by the office did move towards a less restrained line of design and presentation.

Shortly after, in 1991, Hosper set up his own firm, and the possibilities offered by the computer began to affect the presentation of his drawings. It was the end of the Rapidograph technical drawing pen, the end of the adhesive films, the end of the drawing board. The computer also led to specialization and a division of labor that had been uncharacteristic of the profession until that time. Presentation work became a separate function in almost every office. Almost at the same time—probably inspired by the possibilities given by computers and advanced graphic software programs—presentation styles were developed by many firms as a symbolic signature, a way of differentiating oneself, a sort of logo.

Technically speaking, the production of the presentation itself was no longer something at which landscape architects were expected to excel, but Hosper nevertheless wanted to be involved in this part of the process. In his own office, he aimed for a presentation style that was characterized first and foremost by clear line work and vivid color [14-9]. He disliked the increasing tendency in the world of landscape design to allow the signature of a firm or designer to dominate the subject matter. As far as he was concerned, his practice did not require a totally consistent and recognizable graphic style, an attitude probably derived from Hosper's ideas about design. He stressed the importance of implementing an idea both collaboratively and enthusiastically, which reduced the status of the drawing to that of a practical object.

14-9

Alle Hosper. Beverwijk, The Netherlands, c. 1994.

This project characterized the style of the Hosper office. Working with soft colored pencils to obtain a full chromatic range led, in Hosper's view, to vivid drawings.

Bureau Alle Hosper

14-10

Alle Hosper. Beestenmarkt, Leiden, The Netherlands, 1993. Preliminary sketch.

"This is how it all fits together." The idea behind the sketch was to show that by simplifying the infrastructure, the water and town square could together form a single public space.

Bureau Alle Hosper

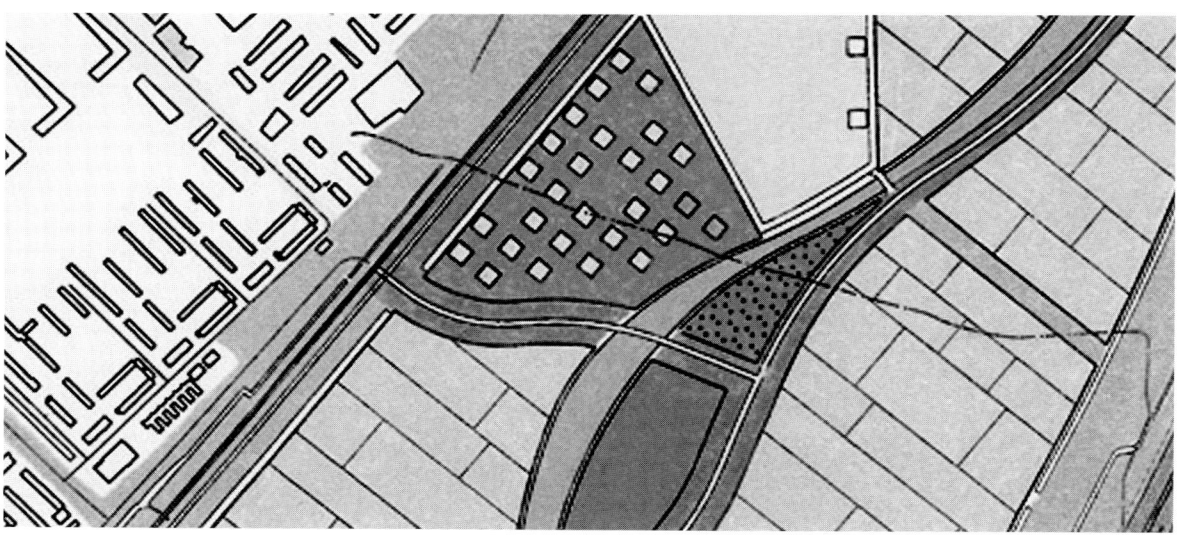

Although he wanted drawings to be attractive and vivid, they were, in fact, primarily a means of communicating the idea of a design as clearly as possible.

SCRIBBLES

Rough sketches, schemes, and drafts provide greater insight into a designer's train of thought—how ideas come into being and how they develop—than the final documents intended for client presentations or publication. Alle Hosper normally sketched the initial drawings himself. Each of his quick scribbles in pencil, felt-tip pen, or chalk is easily recognizable as a "Hosper." These scribbles are interesting because they offer insights into his interpretation of the design program. His sketches for the Beestenmarkt, a square in the center of Leiden, exemplify this stance: They are not well-thought-out propositions, easy to convert into a finished design [14-10]. Rather, they are notes that test his observations of the current situation from one or two different points of view. It is as if the drawings are saying: "This is worth a further look; this could be useful for developing the design." At times, these rough studies even reveal the context in which they were made: at the office, as part of a brainstorming session, or during a discussion with the client.

Above all, Hosper's rough sketches acted as a means of communication. They demonstrate that Hosper's first step while talking to customers or other interested parties was to reach agreement on and generate enthusiasm for a certain line, theme, atmosphere, or concept. Made on the spot, the sketches seem to convey the message: "Look, this is what I mean." For this sort of statement, made in this sort of situation, the traditional drawing methods still seem to be the most effective.

In the late 1980s, the municipality of The Hague started the "De Kern Gezond"

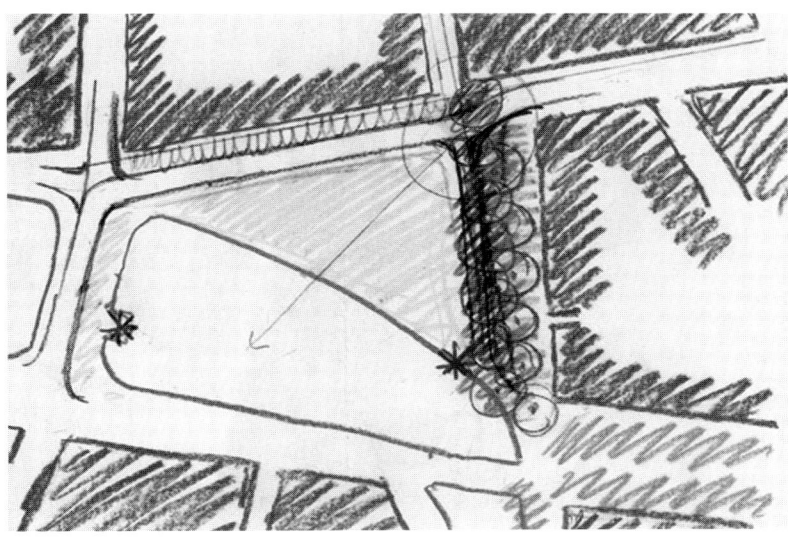

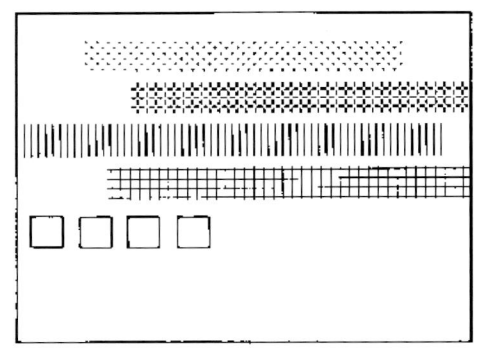 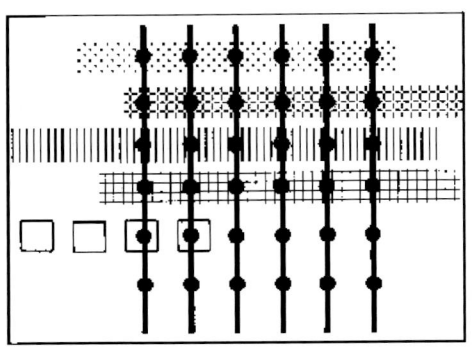

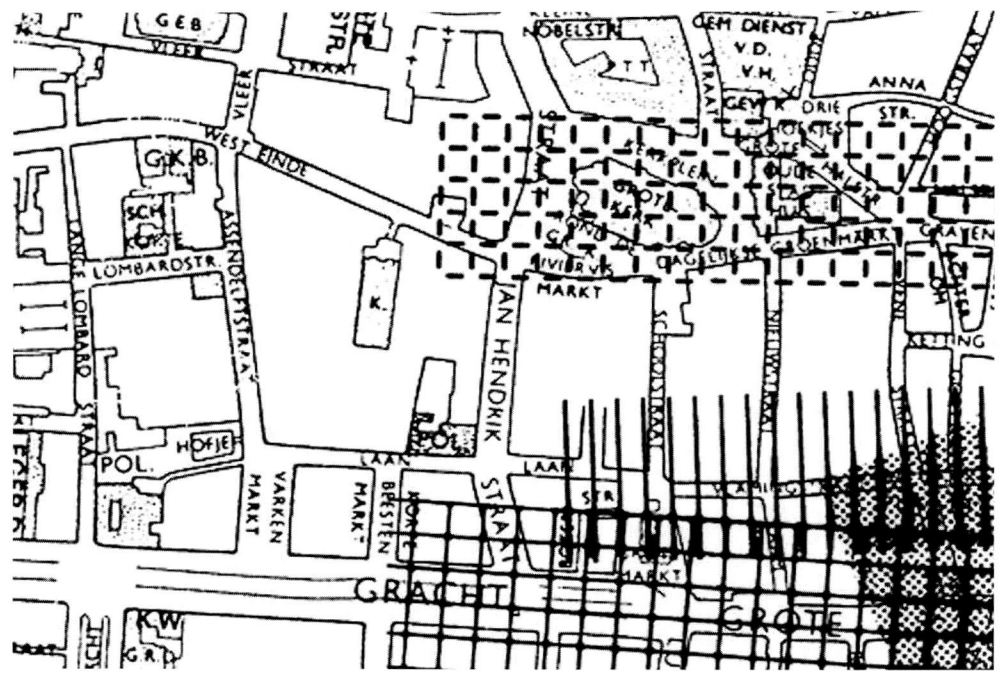

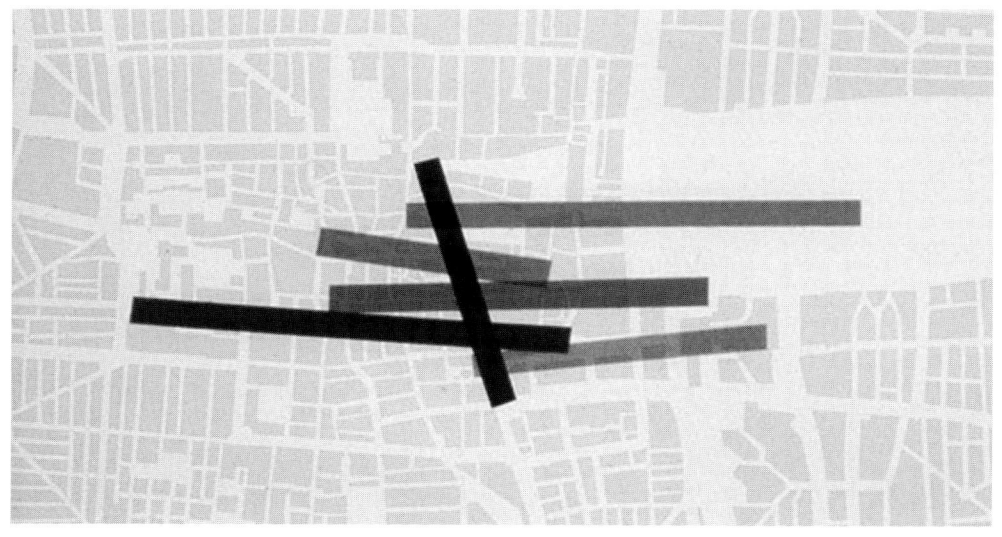

(Healthy to the Core) project, a long-term project that combined a strategic evaluation of all the inner city public spaces with the formulation of very detailed design guidelines. Hosper was invited to chair the initiative which involved many participating municipal departments. Several drawings for this project can be viewed as a summary of Hosper's graphic development, as the entire palette of graphic techniques available in the 1990s appeared in its representations. Colored adhesive films and computer applications together produced a lucid graphic story. Preparatory studies for the project, however, consisted of analytical line drawings with a "technical" feeling similar to that in the earlier mentioned schemes developed for Ameland and Terschelling. As the De Kern Gezond project progressed, these schemes seemed to undergo a graphic metamorphosis that was perhaps directly related to recent technical innovations, thus allowing these two graphic styles to gradually converge [14-11A, 14-11B, 14-11C]. Sadly, the preliminary rough sketches, comparable to the ones made in Leiden, have been lost, but the nature of the project and Hosper's role in it suggest these sketches must have existed.

In summary, one can say that Alle Hosper eagerly tried and tested new drawing materials throughout his many years of professional practice. While open to these fresh techniques, he maintained a belief that drawing methods should serve primarily as vehicles for communication, and he continued to use traditional techniques as well as the new. After all, they had proved their worth over many years. As a result, through his work we can trace the technical development of graphic media, their influence on the products of landscape architecture, and the social context in which these products were made.

14-11A, 14-11B, 14-11C
De Kern Gezond, The Hague,
The Netherlands, late 1980s.

From outline to draft: the graphic development of the project.

NOTES

1 Noël van Dooren, Herma Hekkema, Rob van Leeuwen, and Marinke Steenhuis, *Alle Hosper, landschapsarchitect, 1943–1997*, Rotterdam: 010 Publishers, 2003.

2 Interview with Lodewijk Baljon by Anja Guinée as part of the research for an article on Hosper in the Dutch journal *Blauwe Kamer*, 3, 2000.

3 Interview with Ellen Brandes, Kraggenburg, 2000.

4 Ibid.

5 Alle Hosper, *Een praktijktijd bij de afdeling landschapsarchitectuur van het Staatsbosbeheer*, Wageningen, 1968.

6 Interview with Christiaan Zalm, Kraggenburg, 2000.

Thorbjörn Andersson is a practicing landscape architect, historian, and journalist, and a founding editor of *Utblick Landskap*, the Swedish review of landscape architecture. He has authored, co-authored, or edited numerous publications, including *Svensk Trädgårdskonst under fyrahundra år* (Four Hundred Years of Swedish Landscape Architecture) and *Utanför Staden: Parker i Stockholms förorter* (Outside the City: Parks in the Stockholm Suburbs).

Stephen Daniels is a geographer and landscape historian, and Professor of Cultural Geography at the University of Nottingham. His graduate education spanned study at St. Andrews, the University of Wisconsin, and finally a PhD from the University of London. He has been a prolific writer whose co-authored The *Iconography of Landscape* broke new ground in cultural landscape studies. He is also the author of *Fields of Vision: Landscape Imagery and National Identity*, and most recently *Humphry Repton: Landscape Gardening and the Geography of Georgian England*.

Dianne Harris teaches in both the landscape architecture and architecture departments of the University of Illinois, Urbana-Champaign. She holds an undergraduate degree in landscape, a masters of architecture, and a PhD in architectural history with a dissertation on a landscape topic—all from the University of California, Berkeley. She is the co-editor of *Villas and Gardens in Early Modern Italy and France*, and *The Nature of Authority*, drawn from her dissertation on the Lombard villa landscapes depicted by Marc'Antonio Dal Re.

David L. Hays, Associate Professor of Landscape Architecture at the University of Illinois, Urbana-Champaign, first studied fine arts and literature as an undergraduate, then a PhD in art history from Yale University, and thereafter a masters degree in architecture from Princeton University. In 2004, he edited *306090*, volume 7, "Landscape within Architecture."

Kenneth Helphand, Professor of Landscape Architecture at the University of Oregon received his undergraduate education at Brandeis and graduate study in landscape architecture at Harvard University. A former editor of *Landscape Journal* and fellow of the American Society of Landscape Architects, his publications include *Yard, Street Park* and *Colorado: Visions of an America Landscape, Dreaming the Garden*—the first history of landscape architecture in Israel—and most recently, *Defiant Gardens*.

Randolph Thompson Hester, Jr. studied landscape architecture at North Carolina State University in Raleigh, with graduate study thereafter at Harvard University. He is known as a untiring upholder of the values of all segments of society and an ardent environmentalist. He is the author or co-author or co-editor of a number of books including the *Community Design Primer* and the evergreen and highly influential *The Meaning of Gardens.*

Walter Hood is Professor of Landscape Architecture at the University of California, Berkeley, and a practicing landscape architect. He received his undergraduate degree at North Carolina A&T, and masters degrees in architecture and landscape architecture at the University of California, Berkeley. His exhibition catalog *Jazz and Blues Landscapes* was followed by the monograph *Urban Diaries.*

Dorothée Imbert, Associate Professor of Landscape Architecture at Harvard University, received her professional architectural diploma in France, followed by graduate study towards masters degrees in both architecture and landscape architecture at the University of California, Berkeley. She is the author of *The Modernist Garden in France*, which remains the definitive study of the subject, and the co-author of *Garrett Eckbo: Modern Landscapes for Living.*

Laurie Olin first studied architecture at the University of Washington, before turning to landscape architecture. He has taught for almost his entire professional career, principally at the University of Pennsylvania, where he is currently Professor in Practice, and at Harvard University, where he also served as chair. He has authored numerous publications, most recently *Across the Open Field: Essays Drawn from the English Landscape.*

Kirt Rieder completed studies in urban planning at the University of Cincinatti before taking his master of landscape architecture at Harvard. He has worked for Hargreaves Associates for over a decade, involved in significant projects such as the University of Cincinnati campus, the same city's waterfront, and more locally, the planned development and restoration of Crissy Field in San Francisco. In his design work he has instigated the use of a number of innovative representational techniques, among them the give and take between the physical and virtual model.

Marc Treib is Professor Emeritus of Architecture at the University of California, Berkeley, a practicing designer, and a frequent contributor to architecture, landscape and design journals. He has held Fulbright, Guggenheim, and Japan Foundation fellowships, and an Advanced Design Fellowship at the American Academy in Rome. Among his numerous publications are *Modern Landscape Architecture: A Critical Review*; *The Architecture of Landscape, 1940–60*; *Noguchi in Paris: The Unesco Garden*; *Thomas Church, Landscape Architect: Designing a Modern California Landscape*; and *Settings and Stray Paths: Writings on Landscapes and Gardens*.

Chip Sullivan completed both undergraduate and graduate degrees in land-scape architecture at the University of Florida, and then worked for a number of years at Sasaki and Associates, before spending more than a year in Rome as a Fellow of the American Academy. He is currently Professor of Landscape architecture at the University of California, Berkeley, where he teaches design and drawing. His *Drawing the Landscape* has been widely adopted as text in schools across the country; his second book, *Garden and Climate*, was published in 2002.

Noël van Dooren studied landscape architecture at Wageningen University in The Netherlands, and then worked at H+N+S landscape architects from 1992 to 1997. Since then he has maintained an independent landscape practice. He has published on a wide range of subjects, many of which appeared in the Dutch landscape architecture journal *Blauwe Kamer*, where he served as an editorial board member from 1994 to 2004. He currently heads the Department of Landscape Architecture at the Academy of Architecture in Amsterdam.

Peter Walker, a celebrated landscape architect, took his undergraduate education at Berkeley and passed through Illinois on his way to a masters at Harvard. A fellow of the American Society of Landscape Architects, he has been the recipient of design awards too numerous to count. Key projects include Foothill College and the Weyerhaeuser offices with Sasaki Walker, and the IBM campus and town center at Solana, Texas. He has also chaired the landscape departments at both Harvard and Berkeley, and is the co-author of *Invisible Gardens*, one of the handful of significant studies of landscape architecture of the twentieth century.